INSTRUCTOR'S MANU
Small Animal Care and Management
Third Edition

INSTRUCTOR'S MANUAL TO ACCOMPANY
Small Animal Care and Management
Third Edition

Dean M. Warren

DELMAR
CENGAGE Learning

Australia • Brazil • Japan • Korea • Mexico • Singapore • Spain • United Kingdom • United States

Title: Instructor's Manual to Accompany Small Animal Care and Management, Third Edition
Author(s): Dean M. Warren

Vice President, Career and Professional Editorial:
  Dave Garza
Director of Learning Solutions: Matthew Kane
Managing Editor: Marah Bellegarde
Product Manager: Christina Gifford
Editorial Assistant: Scott Royael
Vice President, Career and Professional Marketing:
  Jennifer McAvey
Marketing Director: Deborah Yarnell
Marketing Manager: Erin Brennan
Marketing Coordinator: Jonathan Sheehan
Production Director: Carolyn Miller
Production Manager: Andrew Crouth
Content Project Manager: Anne Sherman
Art Director: David Arsenault

© 2010, 2002 Delmar, Cengage Learning

ALL RIGHTS RESERVED. No part of this work covered by the copyright herein may be reproduced, transmitted, stored, or used in any form or by any means graphic, electronic, or mechanical, including but not limited to photocopying, recording, scanning, digitizing, taping, Web distribution, information networks, or information storage and retrieval systems, except as permitted under Section 107 or 108 of the 1976 United States Copyright Act, without the prior written permission of the publisher.

> For product information and technology assistance, contact us at
> **Cengage Learning Customer & Sales Support, 1-800-354-9706**
>
> For permission to use material from this text or product,
> submit all requests online at **www.cengage.com/permissions**
> Further permissions questions can be e-mailed to
> **permissionrequest@cengage.com**

Library of Congress Control Number: **2009921854**

ISBN-13: 978-1-4180-4106-9

ISBN-10: 1-4180-4106-8

Delmar
5 Maxwell Drive
Clifton Park, NY 12065-2919
USA

Cengage Learning is a leading provider of customized learning solutions with office locations around the globe, including Singapore, the United Kingdom, Australia, Mexico, Brazil, and Japan. Locate your local office at: **international.cengage.com/region**

Cengage Learning products are represented in Canada by Nelson Education, Ltd.

To learn more about Delmar, visit **www.cengage.com/delmar**

Purchase any of our products at your local college store or at our preferred online store **www.ichapters.com**

NOTICE TO THE READER
Publisher does not warrant or guarantee any of the products described herein or perform any independent analysis in connection with any of the product information contained herein. Publisher does not assume, and expressly disclaims, any obligation to obtain and include information other than that provided to it by the manufacturer. The reader is expressly warned to consider and adopt all safety precautions that might be indicated by the activities described herein and to avoid all potential hazards. By following the instructions contained herein, the reader willingly assumes all risks in connection with such instructions. The publisher makes no representations or warranties of any kind, including but not limited to, the warranties of fitness for particular purpose or merchantability, nor are any such representations implied with respect to the material set forth herein, and the publisher takes no responsibility with respect to such material. The publisher shall not be liable for any special, consequential, or exemplary damages resulting, in whole or part, from the readers' use of, or reliance upon, this material.

Printed in the United States of America
1 2 3 XXX 12 11 10 09

CONTENTS

SECTION 1
- **CHAPTER 1** Introduction to Small Animal Care 2
- **CHAPTER 2** Safety 3
- **CHAPTER 3** Small Animals as Pets 5
- **CHAPTER 4** Animal Rights and Animal Welfare 8
- **CHAPTER 5** Careers in Small Animal Care 9
- **CHAPTER 6** Nutrition and Digestive Systems 13

SECTION 2
- **CHAPTER 7** Dogs 18
- **CHAPTER 8** Cats 24
- **CHAPTER 9** Rabbits 27
- **CHAPTER 10** Hamsters 30
- **CHAPTER 11** Gerbils 32
- **CHAPTER 12** Rats 34
- **CHAPTER 13** Mice 35
- **CHAPTER 14** Guinea Pigs 38
- **CHAPTER 15** Chinchillas 40
- **CHAPTER 16** Ferrets 42
- **CHAPTER 17** Amphibians 44
- **CHAPTER 18** Reptiles 46
- **CHAPTER 19** Birds 51
- **CHAPTER 20** Fish 54
- **CHAPTER 21** Hedgehogs 59
- **CHAPTER 22** Sugar Gliders 61

ANSWER KEY FOR STUDENT WORKBOOK 63

SECTION 1

CHAPTER 1

Introduction to Small Animal Care

DISCUSSION QUESTIONS AND ANSWERS

1. **When did life on earth begin according to scientific data? Please support your response.**
   The oldest direct traces of life on earth date back 3.4 to 3.5 billion years. In rocks that age in Australia and Southern Africa, geologists have found stromatolites, layered structures created through the activity of primitive algae and bacteria. Other Australian rocks of similar age provide even more direct evidence of ancient life. Sections of these rocks, known as cherts, show fossilized remains of blue-green algae.

   Rocks also reveal even more distant, indirect traces of life. Living things use particular isotopes (atomic forms) of the element carbon preferentially. The mix of carbon isotopes detected in rocks from Greenland more than 3.8 billion years ago shows evidence of life on earth.

2. **When did plants first appear on earth? How do we know this?**
   The first land plants were believed to have become established about 420 million years ago during the Ordovician period. This is also the period of time when invertebrates, such as arthropods and worms, appeared on land. These first terrestrial invertebrates fed on decaying plant material.

3. **Why weren't plants the first life to appear on earth?**
   Some 2.2 billion years ago, free oxygen was present in the atmosphere. Living things used this reactive substance, not decaying plant material, in biochemical functions of their cells. The free oxygen in the atmosphere also produced a layer of ozone, which filters out the ultraviolet light from the sun that is harmful to life.

4. **During which time period were dinosaurs most abundant?**
   Dinosaurs were most abundant during the Jurassic period, 135 million years ago.

5. **During which time period did the placental mammals appear on earth?**
   During the Paleocene and Eocene epochs, 40 to 70 million years ago, placental mammals evolved, dispersed, and adapted to new environments.

6. **Which animals were first to be domesticated? Why?**
   Dogs are believed to be the first animals domesticated, and there are two main theories why. The first theory is that dogs may have evolved from a canid ancestor similar to the pariah dogs that populate many underdeveloped countries today. The second theory is that dogs are domesticated wolves. In this theory, wolf pups were hand-raised by humans.

7. **What agency has responsibility for regulating the care and management of small animals?**
   The United States Department of Agriculture has the responsibility for regulating the care and management of small animals.

CHAPTER 2
Safety

DISCUSSION QUESTIONS AND ANSWERS

1. **What is a parasite?**
   Parasites are organisms that live on or within another organism or host and derive their sustenance (food or nourishment) from the host.

2. **What is a host? What is an intermediate host?**
   A host is the organism on which or within which the parasite lives and from which it derives its sustenance (food or nourishment).
   An intermediate host is a host that a parasitic organism lives on or in during an immature stage.

3. **What is a zoonosis?**
   A zoonosis is a disease that can be transmitted from animals to humans.

4. **What is the life cycle of the deer tick and the black-legged tick?**
   On the East Coast, white-footed mice serve as a reservoir (an immune host) for ticks carrying the *Borrelia* organism. These ticks have a 2-year life cycle. They lay their eggs in the spring, and the larvae that emerge feed on white-footed mice. The larvae then remain dormant during the winter and develop into nymphs the following spring. The nymphs feed for 3 or 4 days on white-footed mice and then move to their preferred host, the white-tailed deer.

5. **What is the life cycle of the western black-legged tick?**
   The cycle in Question 4 is followed by the Western black-legged tick except that the dusty-footed woodrat, a common California rodent, is the reservoir for the *Borrelia* organism in the West.

6. **Who is at greatest risk of tick infection? What safety precautions should they take?**
   Pet owners who allow their pets outside and humane shelter workers are probably at greatest risk of tick infection. Such people should wear rubber gloves and take necessary precautions. Other people at high risk are trappers, hunters, hikers, and others who venture into grassy and woody areas. When in areas where ticks may be founds one should wear long pants, long-sleeved shirts, and long socks. Insect repelient is also recommended.

7. **Why are the tapeworm species discussed in this chapter so hard to identify?**
   Echinococcus multilocularis and *Echinococcus granuosus* are difficult to identify because the eggs are identical to the *Taenia* tapeworm species that is common to dogs and cats.

8. **Why are children and the elderly more at risk from some of the diseases and injuries covered in this chapter?**
   Children and the elderly are at a greater risk because they have a lower resistance to some of the disease-causing organisms. Children are also at risk because they play with animals and play in areas that can become contaminated by animals.

9. **Why should a pregnant woman avoid handling a cat's litter box?**
   Toxoplasmosis is a disease produced by infection of the parasite *Toxoplasma gondii*. It can be carried by several different animal species but is usually spread to humans through cat feces or contaminated litter. Pregnant women should not handle cat litter boxes because toxoplasmosis can cause miscarriage, premature births, and blindness in the unborn child.

10. **Why is early diagnosis and treatment important in cases of Rocky Mountain spotted fever?**
    The organism causing Rocky Mountain spotted fever multiplies in the cells of the small peripheral blood vessels. Fever, headache, and skin rash are symptoms of Rocky Mountain spotted fever. Early diagnosis and treatment with antibiotics is important, because the disease can cause death when not treated.

11. **What determines whether someone should be given immune gamma globulin injections for rabies?**
    Whether to treat a patient for rabies will depend on the following criteria:
    - the species of animal that caused the bite
    - the frequency of rabies in the community
    - the circumstances surrounding the bite (Was the animal provoked or was it an unprovoked attack?)
    - the behavior of the biting animal
    - whether the animal can be quarantined and observed
    - whether the animal's head (in the case of a wild animal) can be sent in for laboratory examination of the brain

12. **What are some safety guidelines to follow when working with small animals?**
    The following are some general guidelines to follow for safety in the workplace or school lab:
    a. Always wear protective clothing and equipment when the job requires it.
    b. Always wash protective clothing and equipment after use to prevent contamination.
    c. Wash hands and face after completing a job to make sure all chemical residue is removed.
    d. If required, shower after completing a job so that chemical residue is completely removed from the body.
    e. Wash hands frequently while working with animals, especially if working with different species and in different areas. This prevents contamination to other animals and also prevents self-contamination.
    f. Keep hands away from the mouth, eyes, and face when working with chemicals and animals to prevent self-contamination.
    g. Do not consume food or drinks in areas where contamination could occur, and do not store these items in areas where contamination could occur.
    h. Remove uniforms, lab coats, and coveralls when leaving an area that could be contaminated.
    i. Never wash uniforms lab coats, or coveralls with regular clothing.
    j. Make sure all containers are correctly labeled to prevent misuse of chemicals.
    k. Dispose of all chemicals and their containers according to proper procedure or instructions on the labels.
    l. Students and small animal workers should be instructed in the proper methods of handling small animals.
    m. First-aid kits should be kept in the work area or instructional area, and workers and students should be made aware of the location of first-aid kits.

CHAPTER 3

Small Animals as Pets

DISCUSSION QUESTIONS AND ANWERS

1. **Are there unwanted animals in your area? What happens to them? Does a problem exist with unwanted animals? What should be done to correct the problem?**
   Answers to these questions will vary with communities. This is an opportunity to discuss the over population problems, unwanted animals, and what happens to these animals.

2. **What are the questions one should ask before obtaining a pet?**
   Before obtaining a pet, a person should consider the following questions carefully:
   a. Where should I obtain a pet?
   b. How much space do I have for a pet?
   c. What kind of animal does my lifestyle allow?
   d. How much will the animal cost?
   e. What will the future bring?
   f. Does everyone in the family want a pet?
   g. What kind of personality do I have?
   h. Is this animal a fad or status symbol?
   i. What am I going to use the animal for?

3. **What are four sources from which you can obtain pets? What are the advantages and disadvantages of each source?**
   a. Pet shops
      Advantages: conveniently located, full line of pet care products, counseling may be offered
      Disadvantages: consumer may not know where animals come from or conditions under which they were raised or shipped and may not have any knowledge of pedigree; more expensive
   b. Purebred breeders
      Advantages: reputation is important for their success, usually guarantee satisfaction, cheaper; consumer can see condition of kennel and observe parents, ancestry is known
      Disadvantages: may take more time to locate suitable animal; may have to travel to pick up animal
   c. Animal shelters
      Advantages: less expensive, saving an animal's life
      Disadvantages: many are not purebred, no knowledge of parents, traits of parents cannot be observed
   d. Friends and neighbors
      Advantages: easy way to obtain a pet, often cheapest source (may be free)
      Disadvantages: sire usually unknown, only traits of dam can be observed

4. **What are some of the various lifestyles that people have? What pets might best fit those lifestyles?**
   Following are some examples of answers that may be given. There are other suitable answers.
   - Small apartments—small dogs, quiet cats
   - Large apartments—larger dogs, more active cats
   - Always busy—caged animals such as hamsters, fish
   - A lot of free time—long-haired animals, birds
   - Active lifestyle—larger dogs and dogs needing lots of exercise
   - Inactive lifestyle—small, lap dogs; inactive cats and caged animals

5. **What are some other costs of owning a pet besides the initial purchase price?**
   - Cage and equipment
   - Toys
   - Food
   - Grooming and/or grooming supplies
   - Taxes and licenses
   - Veterinary bills

6. **What lessons can children learn from having a pet? Are there other lessons not listed in the text?**
   - Responsibility—feeding and caring for the needs of a pet
   - Social skills—caring for and understanding a pet
   - Respect and compassion
   - Grief and coping with loss

7. **Do animals have personalities? What are some of the personalities you have observed?**
   Animals have personalities just like people. Dogs vary greatly in their personalities; some are docile, whereas others are active and playful. Cats generally are less active than dogs; some are quiet and lay around, but others are more active, need lots of attention, and are more vocal. Some birds are vocal and need attention, whereas others are quiet.

8. **What are some animal fads or status animals of recent years?**
   The following animals have been the object of fads of status: Shar-Pei, Afghan, Dalmations, Chihuahuas, Poodles, Pot-bellied pigs, and Cheetahs and other exotic cats.

9. **What do we mean by responsible pet ownership?**
   Responsible pet ownership means providing for all the needs of a pet, including the following:
   - making sure the animal has food and water
   - licensing and paying required taxes
   - obeying leash laws, fence laws, and other control laws
   - not allowing the animal to roam the neighborhood
   - seeing that the animal gets veterinary attention as needed
   - having the animal receive regular vaccinations
   - having the animal spayed or neutered to prevent unwanted pregnancies

10. **What do we mean by the terms *spay* and *neuter*? Should we have our pets spayed or neutered?**
    Spaying is performed on females; the ovaries and uterus are removed. This prevents the female from coming into heat and from becoming pregnant.

    Neutering is performed on males; the testicles are removed. This prevents the male from producing sperm and impregnating a female.

    Each year, approximately 27 million dogs and cats are born in the United States; 50 million are euthanized each year as unwanted or abandoned. To solve the overpopulation and unwanted animal problem, the number of puppies and kittens born must be reduced. Adults and children need to be educated and take responsibility for controlling the breeding of dogs and cats.

11. **What are the benefits of pets to the elderly?**
    Pets can help the elderly live longer, happier lives by helping to lower blood pressure and helping the owners recover from heart attacks and other serious illnesses.
    Pets can help relieve the stress of major illnesses, death in the family, or divorce.
    Pets can decrease loneliness and give the elderly an opportunity to be needed.
    Pets can stimulate the elderly to exercise.
    The elderly can feel a sense of responsibility by caring for and providing attention to a pet.

12. **What are some of the therapeutic uses of pets?**
    Pets provide a great benefit to people with handicaps. Dogs can be vital in both a practical and psychological way for people who are hearing impaired, vision impaired, or confined to wheelchairs or beds. Pets can help elicit responses from severely withdrawn psychiatric patients.
    Pets are good for latchkey children who would otherwise come home to empty houses after school and for children who tend to be picked on by their classmates.

13. **What is euthanasia, and why should it be considered by a pet owner? How would you counsel someone who is dealing with the loss of a pet?**
    Euthanasia is to induce the death of an animal quickly and painlessly, putting it to sleep. Euthanasia may be the kindest thing you can do for a pet that is severely sick of injured.
    Grieving over the loss of a pet is natural and common. When people are mourning the loss of a pet, they should be allowed to express their feelings.

CHAPTER 4

Animal Rights and Animal Welfare

DISCUSSION QUESTIONS AND ANSWERS

1. **What was the title and author of the book that aroused public indignation over the welfare of animals in Great Britain?**
   The book was *Animal Machines: The New Factory Farming Industry* by Ruth Harrison.

2. **What is a "factory farm"?**
   A "factory farm" is a farm where chickens are kept in cages and veal calves in crates.

3. **Who is considered the founder of the modern animal rights movement, and what was the name of his publication?**
   Peter Singer is considered the founder, and his publication was titled *Animal Liberation*.

4. **What do the terms *animal rights* and *animal welfare* mean?**
   Animal rights is the position that animals should not be exploited. Animal rights advocates believe that animals should not be used for food, clothing, entertainment, medical research, or product testing. This also includes the use of animals in rodeos, zoos, circuses, and even as pets. They believe it is ethically, morally, and inherently wrong to use animals for human purposes under any condition.
   Animal welfare is the position that animals should be treated humanely. This includes proper housing, nutrition, disease prevention and treatment, responsible care, handling, and humane euthanasia or slaughter. Animal welfare advocates believe that animals can be used for human purposes but that they should be treated so that discomfort is kept to a minimum.

5. **Do animals have rights?***

6. **Should animals be used for food?***

7. **Should animals be used for experimentation?***

8. **Should animals be used for other purposes, that is, for hunting or trapping, for entertainment purposes in zoos and exhibits, for rodeos, or even for pets?***

*The answers to questions 5, 6, 7, and 8 will vary depending on the views of the students. This is an opportunity for students to express their views and for class discussion on animal rights and animal welfare issues.

CHAPTER 5

Careers in Small Animal Care

DISCUSSION QUESTIONS AND ANSWERS

1. **How large is the animal care industry? How important is it in your area?**
   The pet industry in the United States is a rapidly growing segment of today's business world. Americans spend in excess of $40.8 billion a year on their pets. Veterinary expenses exceed $24.5 billion annually, and pet food manufacturers produce some $15.4 billion in sales. The pet care industry is just one area of small animal care.
   Answers will vary on the second part of the question because of differences in communities.

2. **What are the duties of the workers in the various small animal careers?**
   Pet care workers provide a wide variety of services for the owners of small animals.
   Animal caretakers complete tasks that range from the day-to-day care for healthy animals to the care for sick, injured, or aging animals.
   Animal breeders and technicians help breed, raise, and market various animals.
   Aquarists work for aquariums, oceanariums, and marine research institutes. They are responsible for the maintenance of aquatic exhibits.
   Naturalists educate the public about the environment and maintain the natural environment on land dedicated to wilderness populations.
   Animal handlers care for, train, and study animals in places like zoos, parks, and research facilities.
   Zoo and aquarium curators and directors coordinate the business affairs of zoos and aquariums.
   Animal trainers teach animals to obey commands, compete in shows or races, and perform tricks to entertain audiences, protect property, or act as guides for the disabled.
   Pet groomers comb, cut, trim, and shape the fur of dogs and cats. They bathe pets and use special solutions to keep them free from ticks, fleas, and other pests. They often brush the pets and trim their hair and nails.
   Veterinarians, also called doctors of veterinary medicine, diagnose and control animal diseases, treat sick and injured animals medically and surgically, prevent transmission of animal diseases, and advise owners on proper case of pets and livestock.
   Veterinary technicians assist veterinarians and other members of the veterinary staff in diagnosing and treating animals for injuries, illness, and routine veterinary needs, such as laboratory testing procedures.
   Biologists study the origin, development, anatomy, function, distribution, and other basic principles of living organisms. They do research in many specialties that advance the fields of medicine, agriculture, and industry.
   Zookeepers provide day-to-day care for zoo animals, as well as interact with visitors, conduct formal and informal educational presentations, and help with trainers and research. Generally, their job is to make sure that the animals are healthy and ready to be exhibited. They may also repair fences, exercise the animals, and make sure that visitors do not do anything to harm the animals.
   Zoologists are biologists who study animals.

3. **What are some of the different working conditions one might encounter?**
Animal caretakers may work indoors or outdoors. They must get used to animals and should not mind cleaning up after them. Their work may involve lifting heavy animals and equipment. They may have to drive a station wagon, van, or light delivery truck to pick up and deliver pets. Their work involves dealing with people, so workers should be friendly and courteous. Many pet care workers work 40 hours a week. Some must also work or be on call evenings and weekends.

Animal breeders and technicians work in all types of facilities, from barns and pens to private homes or housing facilities.

Zoo administrators work long and exhausting hours. They may take paperwork home with them, stay at the zoo through the night nursing a sick animal, or get up in the middle of the night to meet a pair of rhinos arriving at the airport. They must pay close attention to detail and have great patience. Animal curators, veterinarians, and zoologists have the satisfaction of working with and learning from the animals. The director works under a great deal of pressure.

The working hours for animal trainers vary considerably, depending on the type of animal, the performance schedule, and whether travel is involved. For some trainers, such as those who work with show horses, hours can be long and quite irregular. Except in warm climates, animal shows are seasonal, running from April or May through mid-autumn. During this time, much of the work is conducted outdoors. In winter, trainers work indoors, preparing for warm-weather shows. Trainers of aquatic mammals, such as dolphins and seals, must feel at ease working around water. The physical strength required depends on the animal involved, and animal trainers usually need greater-than-average agility.

Working conditions for pet groomers can vary greatly, depending on the location and type of employment. Many salons and pet shops are clean and well lit, with modern equipment and clean surroundings. Others may be cramped and dark. Groomers need to be careful while on the job, especially when handling flea and tick killers, which are toxic to humans. When working with any sort of animal, a person may encounter bites, scratches, strong odors, fleas, and other insects. They may also have to deal with sick or bad-tempered animals. Groomers who are self employed can work out of their homes. Many people covert their garages into work areas. Some groomers buy vans and convert them into grooming shops. They can drive to the homes of the pets, which many owners find convenient. Groomers usually work a 40-hour week and may have to work evenings or weekends. If they work any overtime, they are compensated for it. Those who own their own shops or work out of their homes, like other self-employed people, work very long hours and can have irregular schedules. Other groomers may work only part-time. Groomers are on their feet much of the day, and their work can get tiring when they have to lift and restrain large animals.

4. **What are the job requirements for the various animal care careers?**
Many of the pet care worker positions are entry level and require no special educational requirements. Many employers hire high school students and high school graduates. Courses in biology, science, and animal husbandry will help prepare students for pet care jobs. Business courses such as bookkeeping will help prepare students for running a business.

A high school education and 4-year college degree are usually required to become a zookeeper. Science courses in biology are helpful too. To work in zoos that are operated by the government, one may have to pass a civil service examination. Animal caretakers must be patient and enjoy working with animals.

Persons wanting to be animal trainers should like animals and have a genuine interest in working with them. Establishments that hire trainers often require previous experience as an animal keeper or aquarist because proper care and feeding of the animals is an essential part of a trainer's responsibilities. Some positions require college degrees. Zoo and aquarium animal trainers usually must have bachelor's degrees in a field related to animal management or animal physiology.

To get the Doctor of Veterinary Medicine (DVM of VMD) degree, one must have graduated from one of the 28 accredited schools of veterinary medicine in the United States. To become licensed to practice veterinary medicine, one must also pass the state's oral and written licensing examination. One must have at least 2 years of undergraduate training at a college or university before one can apply for admission to a veterinary college. Many students earn a bachelor's degree before they apply for admission.

To become a veterinary technician, one needs to complete a 2- to 4-year college-based program leading to the associate degree from a community or technical college whose program has been accredited by the American Veterinary Medical Association. A high school diploma is an essential requirement for admission to such a program. Courses in algebra, chemistry, biology, and English are essential.

All biologists have earned undergraduate degrees in science. Most go on to complete master's and doctorate degrees in their areas of interest. A master's degree and a doctorate are essential for any biologist who wishes to conduct serious research, publish in scholarly journals, and obtain a faculty position at a college or university.

5. **What are some background requirements that would be helpful when applying for a job in small animal care?**

Students interested in pet care worker positions should have some experience with animals, and knowing the needs and habits of animals is an asset.

Zookeepers can gain experience working with animals by getting a job with a local veterinarian, pet shop, or animal shelter.

Most trainers begin their careers as keepers and gain on-the-job experience in evaluating the disposition, intelligence, and trainability of the animals they look after. At the same time, they learn to develop a rapport with the animals. Although previous training experience may give job applicants an advantage in being hired, they still will be expected to spend time caring for animals before advancing to positions as trainers.

Part-time or volunteer work in animal shelters, pet shops, or veterinary offices offers would-be trainers a chance to become accustomed to working with animals and to discover whether they have the aptitude for it. Experience can be acquired in summer jobs as caretakers at zoos, aquariums, museums that feature live animal shows, and amusement parks. For those with a special interest in horses, racing or riding stables can provide experience.

Pet groomers can obtain experience by working in pet shops and kennels.

6. **What requirements are necessary for advancement in the various careers?**

Pet care workers advance by becoming supervisors in large animal hospitals, kennels, or pet shops. Some experienced workers start their own kennels or pet shops. Pet care workers can increase their earnings by developing skills in special fields such as grooming, training, or breeding animals.

Zookeepers who have a great deal of experience may advance to positions such as head keepers. The employment outlook for animal caretakers is good because zoos are becoming more popular.

Most establishments have small staffs of animal trainers, which means that the opportunities for advancement are limited; the progression is from animal keeper to animal trainer. A trainer who directs or supervises may be designated head animal trainer or senior animal trainer.

Some animal trainers go into business for themselves and, if successful, hire other trainers to work for them. Others become agents for animal acts.

Pet groomers who work for other people may advance to more responsible positions, such as office manager or dog trainer. Dog groomers who open their own shops may become successful enough to expand or open area franchises.

Veterinarians who are in private practice advance by expanding their practices and by developing a good reputation in their communities. Some veterinarians who work for large organizations are promoted to supervisory of management positions.

With experience and continuing education, veterinary technicians can expect to advance to greater levels of responsibility. In some settings, this simply means performing more difficult or sophisticated procedures, or it could mean that the technicians take on supervisory or other administrative responsibilities.

There are many possibilities for advancement in the field of biology, especially for those who have a doctorate degree. Job opportunities for biologists are expected to increase as fast as average through the year 2012, although competition will be stiff for some positions. Biologists with an advanced degree will be best qualified for the most lucrative and challenging jobs, although this varies by specialty, with genetic, cellular, and biochemical research showing the most promise.

7. **What are some high school courses that would be helpful for students to take that might prepare them for a career in small animal care?**
   - biology
   - animal husbandry
   - zoology
   - nutrition
   - animal hygiene
   - chemistry
   - bookkeeping
   - genetics
   - marketing
   - business

CHAPTER 6

Nutrition and Digestive Systems

DISCUSSION QUESTIONS AND ANSWERS

1. **Define the terms *nutrient* and *nutrition*.**
   Nutrient refers to a single class of foods or group of foods of the same general chemical composition that supports animal life. *Nutrition* refers to the animal receiving a proper and balanced food and water ration so that it can grow, maintain its body, reproduce, and supply or produce the things we expect from it.

2. **Define the term *biochemical reaction*. What are the biochemical reactions that take place in the animal's body?**
   A biochemical reaction is a chemical reaction that takes place in the cells of plants and animals. These chemical reactions include respiration, digestion, and assimilation.

3. **What are the nutrient groups, and what is their importance in the animal's diet?**
   Water is in every cell of the animal and is necessary for the following:
   a. supporting biochemical reactions in the body, including respiration, digestion, and assimilation
   b. transporting other nutrients
   c. helping maintain body temperature
   d. helping to give the body its form
   e. carrying waste from the body

   Proteins are needed for the following:
   a. developing and repairing body organs and tissues, such as muscles, nerves, skin, hair, hooves, and feathers
   b. producing milk, wool, and eggs
   c. developing the fetus
   d. building material for enzymes and hormones
   e. developing antibodies
   f. transmitting DNA

   Carbohydrates supply energy needed for the following:
   a. supporting bodily functions, such as breathing, digestion, and exercising
   b. producing heat to keep the body warm
   c. storing fat

   Fats are essential in the diet for the following:
   a. providing energy
   b. aiding in the absorption of fat-soluble vitamins
   c. providing fatty acids

   Vitamins are necessary for specific biochemical reactions such as the following:
   a. regulating digestion, absorption, and metabolism
   b. developing normal vision, bone, and such external body coverings as hair and feathers
   c. regulating body glands
   d. forming new cells

14 Small Animal Care and Management

    e. fighting disease and strengthening the immune system
    f. developing and maintaining the nervous system

Minerals are primarily needed to supply the materials for building the skeleton and producing body regulators such as enzymes and hormones.

4. **List the various vitamins and explain their importance in the animal's body.**

   Fat-soluble vitamins include A, D, E, and K.
   Water-soluble vitamins include C and the B-complex vitamins.
   Vitamin A is associated with good vision, respiration, digestion, and reproduction.
   Vitamin D is associated with calcium and phosphorus in the body.
   Vitamin E is important for successful reproduction.
   Vitamin K is necessary for maintenance of normal blood coagulation.
   Vitamin C, or ascorbic acid, is synthesized by most animals and is not a consideration in feeding rations. Guinea pigs and primates must receive dietary vitamin C to prevent scurvy.
   Vitamin B_1, or thiamine, is necessary for normal metabolism of carbohydrates. It is needed to prevent a deficiency disease called polyneuritis or beri-beri.
   Vitamin B_2, or riboflavin, is necessary for normal bone development.
   Niacin is necessary for digestion and growth.
   Pantothenic acid is necessary for proper growth and nerve development.
   Vitamin B_{12}, or cyanocobalamin, is necessary for normal growth, reproduction, and blood formation.
   Choline functions in the transportation and metabolism of fatty acids.
   Folic acid, or folacin, is required for normal cell development and is essential in certain biochemical reactions.
   Biotin is important in carbohydrate and fat metabolism.
   Vitamin B_6, or pyridoxine, helps build proteins and is essential in the metabolism of carbohydrates.

5. **List the various minerals necessary in the diet of an animal. What is the importance role each plays in the diet of the animal?**

   Calcium is essential for bone, teeth, and eggshell formation; normal blood coagulation; and milk production.
   Phosphorus is essential for the formation of bones, teeth, and body fluids and is required for metabolism, cell respiration, enzyme-based reactions, and normal reproduction.
   Potassium is required for many body functions, such as maintaining osmotic pressure, acid–base balance, activating enzymes, regulating neuromuscular activity, helping to regulate heartbeat, and digestion.
   Sodium is the chief cation of blood plasma and other extracellular fluids and plays a role in the transmission of nerve impulses and in the absorption of sugars and amino acids from the digestive tract.
   Chlorine is associated with sodium and potassium in acid–base relationships and osmotic regulation.
   Sulfur is a component of cartilage, bone, tendons, and blood vessels and is an important part of the respiratory process.
   Magnesium is necessary in the activation of many enzyme systems, particularly those involved with carbohydrate and lipid metabolism, and for proper functioning of the nervous system.
   Iron is necessary in the formation of hemoglobin.
   Copper is necessary for proper iron absorption, hemoglobin formation, in various enzyme systems, and in the synthesis of keratin for hair and wool growth.
   Iodine is important in the production of thyroxin by the thyroid gland.
   Cobalt is an important compound of the vitamin B_{12} molecule and is necessary in ruminant animals for proper synthesis of vitamin B_{12}.
   Manganese is involved with the enzyme systems that influence estrus, ovulation, fetal development, udder development, milk production, growth, and skeletal development.
   Zinc is needed for the body's immune system, for the manufacture of proteins and genetic material, for the activity of more than 200 enzymes, for stimulating hair growth, for the sense of smell and taste, and for healing wounds.

Molybdenum is a component of the enzyme xanthine oxidase and is important in stimulating action of rumen organisms.

Selenium is necessary for the absorption and utilization of vitamin E.

6. **What is the difference between the digestive system of a ruminant and a nonruminant animal?**
   Ruminant animals differ from nonruminants in several ways:
   a. The ruminant stomach consists of four compartments.
   b. Food material in the ruminant system is acted on by millions of bacteria and microorganisms.
   c. These bacteria and microorganisms transform low-quality protein and some nitrogen compounds into essential amino acids.
   d. The bacteria and microorganisms also aid in the manufacture of needed vitamins, including the B-complex group.
   e. Food material not fully digested in the rumen can be regurgitated in the form of cud. The animal chews on this cud and then swallows it back down into the rumen for further digestion.
   f. The ruminant digestive system can utilize large amounts of roughage.

7. **List the four compartments of the ruminant stomach and explain the function of each.**
   a. The rumen is the largest compartment and makes up about 80 percent of the total capacity of the stomach. The rumen is where the bacteria and microorganisms act on the food material and where digestion takes place.
   b. The reticulum is closely associated with the rumen. The reticulum works with the rumen in formation of the cud for regurgitation. Foreign bodies such as nails or pieces of wire can be held in the reticulum for long periods without causing serious injury.
   c. The omasum removes large amounts of water from the food as it moves from the rumen to the abomasum.
   d. The abomasum functions very similarly to the stomach of single-stomached animals.

8. **How does the digestive system of a horse, rabbit, and chicken differ from a ruminant digestive system? Although they are considered nonruminants, how does the digestive system of a horse, rabbit, and chicken differ from those of most other nonruminants?**
   Horse, rabbits, and chickens are single-stomached animals.
   The horse can consume large amounts of forage but is not considered a ruminant animal. It has a small, single stomach, but unlike other single-stomached animals, it has a large cecum and colon located between the small and large intestines. Bacterial action takes place in the cecum and allows the horse to digest roughage material, but not as efficiently as a true ruminant animal. Horses do not have gallbladders. Bile is secreted into the duodenum directly from the liver.
   The rabbit's digestive system is very similar to that of the horse. It, too, has a large cecum that allows for the utilization of high-quality roughage material; bacteria are present in the cecum and help break down roughage.
   Rabbits and rodents eat their feces. This is referred to as coprophagy and is usually done late at night or early in the morning. Feces that are eaten are usually light green and soft and have not been completely digested.
   Birds, although considered single-stomached animals, have several different organs in their digestive systems. Birds do not have teeth, so no chewing or breaking down of food material takes place in the mouth, although saliva is added here to aid in swallowing. Food material passes down the esophagus, into the crop. The next step is the proventriculus, where gastric acids and enzymes are secreted to begin chemical digestion. The food material then passes to the ventriculus, commonly referred to as the gizzard. This is the largest organ of the bird's digestive system. The primary purpose of the gizzard is to grind and crush food before it enters the small intestine. The gizzard is composed of a horny, lined structure that is heavily muscled. Involuntary muscular action serves to break up the food material, much like teeth action in other animals. Some birds, those that consume seeds and whole grains, are often fed "grit" in the form of crushed granite, oyster shell, or other insoluble material that aids in the grinding and breaking down of coarse material.

Food material then passes on to the small intestine. The first part of the small intestine is the duodenum, where enzymes are secreted from the pancreas to help break down the proteins, starches, and fats. Bile is secreted from the gallbladder to aid in the breakdown of fats. Most of the absorption takes place in the lower section of the small intestine.

Birds have two pouches, called ceca, where the small intestine connects into the large intestine. Although some bacteria are present in these pouches, very little digestion of the fiber is believed to occur here. Food material passes from the small intestine into the large intestine and then into the cloaca. The cloaca serves as a common junction for the bird's digestive, urinary, and reproductive systems.

SECTION 2

CHAPTER 7

Dogs

DISCUSSION QUESTIONS AND ANSWERS

1. **What is the history of the dog? How has the modern dog evolved?**
   The dog originated about 12,000 to 14,000 years ago in Europe and Asia and was domesticated about 10,000 years ago. Many of our modern dogs probably descended directly from the wolf. These animals roamed in packs and probably gradually found their way into human encampments. Humans found that they could depend on the dog to warn of danger, and the dog depended more on humans for food and shelter. Modern dogs evolved as a result of selective breeding for specific purposes and as a result of the environment in which they lived.

   The ancestor of the entire dog family is believed to have been a civetlike animal, *Miacis*, that lived 40 or 50 million years ago. *Miacis* was a small animal with an elongated body and was probably arboreal, or at least seemed to have spent considerable time in the lush forests of its era. Next was the appearance, about 35 million years ago, of two apparently direct descendants of *Miacis*. These were *Daphaenus*, a large, heavy-boned animal with a long tail, and *Hesperocyon*, a small, slender animal that was the forerunner of the long-bodied, coyote-like "bear dogs" that eventually evolved into modern bears. *Hesperocyon* can be considered the "grandfather" of the dog family. *Hesperocyon* retained the long body and short legs of the primitive carnivores, but unlike *Miacis*, *Hesperocyon* spent little of its time in the trees and began to hunt on the ground. Its claws were retractile, enabling it to walk on the ground and climb trees.

   From *Hesperocyon* there evolved two distinct dog types. The first, *Temnocyon*, was an important link in the evolutionary chain that led to the modern hunting dog of Africa, the Cape hunting dog. The second, *Cynodesmus*, is regarded as the ancestor of a large and diversified group of dogs that includes the modern Eurasian wolf and the American dogs, foxes and wolves.

   The animal considered to be the "father" of modern dogs, *Tomarctus*, was directly descended from *Hesperocyon*. *Tomarctus* had a body built for speed and endurance as well as for leaping and differed little in appearance from the modern dog. This was a hunter, an animal geared for the chase, that brought down prey by way of slashing teeth. The modern dog still retains much of *Tomarctus*' anatomical structure and is surpassed in speed only by the cheetah.

   As the evolutionary progress of the Canidae family continued, the progeny of *Tomarctus* developed into the modern dogs, wolves, fox, coyotes, fennecs, and jackals. At the same time, descendants of *Temnocyon* gradually emerged as the Cape hunting dog of today.

   Descending in a direct line from *Tomarctus* are four major lines of dogs: the herd dogs, the hounds and terriers, the Northern and toy dogs, and the guard dogs. It is from these four lines, or groups, that modern dogs are descended. Today, there are seven major groups and more than 400 breeds that have been developed in the past 100 to 250 years.

2. **What are the seven major groups of dogs? How do they differ?**
   a. The sporting group was developed to assist the hunter in the pursuit of game.
   b. The hound group is of two basic types: one type hunts by scent and the other by sight.

c. The terrier group is divided into two subgroups: the long-legged, larger breeds and the short-legged, small breeds. Many of the terriers were developed to go into coal mines to hunt mice and rats or to pursue game into their tunnels and underground dens. The word *terrier* is derived from the Latin word *terra*, which means "earth."
   d. The working dog group was developed to labor or work for humans. They may serve as guard dogs, sled dogs, police dogs, rescue dogs, and messenger dogs.
   e. The herding dog group was developed to aid the livestock herder with various species of livestock.
   f. The toy dog group is so named because of size; some of these dogs are as small as 1.5 pounds. These dogs are very alert and popular as house pets and companions.
   g. The nonsporting group consists of miscellaneous breeds with a wide variety of sizes and characteristics. Basically, these dogs are used as companion dogs.

3. **List the different breeds of dogs. What are the characteristics of the different breeds?**
   For the discussion of the characteristics of the different breeds of dogs, please refer to Tables 7–1 through 7–7.

4. **List the various terms that describe the behavior of the different breeds. What do these terms mean?**
   - aloof—one who holds the head high in a proud, confident manner
   - amiable—mild, easygoing
   - aristocratic—holding the head high and having a certain air of nobility
   - bold—showing no hesitation or fear
   - congenial—easygoing and cooperative
   - constitution—the physical makeup of an animal
   - dignified—an excellent-appearing animal that shows beauty of form and grace
   - disposition—how an animal acts under various circumstances
   - docile—an animal that is tame, easygoing, and obedient
   - feisty—nervous, excitable, and easily agitated or irritated
   - gaiety—being cheerful, happy, or merry
   - impetuous—abrupt, hasty, and rushing headlong into its work without much efficiency
   - incorrigible—unable to be controlled or corrected
   - independent—wanting to have its own way and not wanting to be controlled
   - indolent—lazy or inactive
   - jovial—good natured, happy, easygoing
   - keen—very sensitive, alert, and intense
   - meek—showing weakness and a lack of aggression
   - reserved—shy, cautious, or restrained
   - serene—calm and quiet
   - spirit—showing a lot of pep, vigor, and energy
   - staunchness—strong, steady, steadfast while on point
   - steady—staying on task
   - temperament—the emotional and mental qualities of an individual animal
   - thorough—describes a dog that works every bit of ground and cover
   - timid—showing lack of confidence, somewhat afraid
   - tranquil—quiet and peaceful
   - vain—appearing conceited; overconfident or showing excessive pride
   - versatile—having a wide range of skills
   - vigilant—always alert and watchful
   - wary—showing caution; being somewhat afraid

5. **What aspects should be considered before selecting a breed of dog?**
In selecting a specific breed, the following questions should be considered:
   a. Does one want a large or small breed of dog?
   b. Is an active or quiet breed of dog wanted?
   c. What type of hair should the dog have?
   d. What is the purpose of the dog going to be?
   e. What price will have to be paid?

6. **What are the various feeds available for dogs?**
Three main types of commercial foods are available today: dry, semi-moist, and canned. These differ primarily in their ingredients and the amount of moisture they contain. The main ingredients in dry foods are corn, soybean meal, wheat millings, meat, and bone meal. Ingredients in semi-moist foods are corn, meat by-products, soybean meal, and corn syrup. Canned foods are of two types: (1) a ration type that contains barley, meat by-products, wheat grain, and soy flour, and (2) a meat-type that contains meat by-products, meat, poultry, and soy flour. Dry foods contain about 10 percent moisture, semi-moist contain 30 percent, and canned foods 75 percent.

Their nutrient compositions also differ. Dry foods contain approximately 23 percent protein, 9 percent fat, and 6 percent fiber. Semi-moist foods have about 25 percent protein, 9 percent fat, and 4 percent fiber. The ration-type canned foods contain about 30 percent protein, 16 percent fat, and 8 percent fiber, whereas the meat-type canned foods contain 44 percent protein, 32 percent fat, and 4 percent fiber. Most foods contain about the same amount of energy.

Dry foods have the advantage of being cheaper to purchase, are convenient to use, will not spoil, and help keep the dog's teeth clean.

7. **How does the feeding of pregnant and lactating females, of puppies, and of older dogs differ?**
The amount of ration given to the pregnant female should increase as her weight increases. Her weight normally starts to increase about the fourth week of pregnancy, and just before giving birth, she may be consuming 35 to 50 percent more food than before. She should be fed in three or four evenly spaced meals to avoid the discomfort that one large meal may cause. Modern commercial food provides an adequate and balanced ration during pregnancy, and supplemental feeding should not be needed.

Following whelping, the female will need two to three times as much food as before. This should be given in three or four meals per day. The ration should be the same as has been provided throughout the pregnancy.

Beginning at about 3 weeks of age, the young puppies will begin to lap food from their mother's food bowl, and the owner will need to consider weaning the pups. This can be done gradually, as the puppies begin to consume more solid food. Remove the puppies from the mother during the day, and increase the amount of solid food given to the puppies. Weaning should be complete when the puppies are about 6 weeks of age. The female can then be returned to her regular maintenance diet.

As mentioned in the previous paragraph, puppies will begin to lap solid food at about 3 weeks of age. Puppies can be started on solid food by giving them canned puppy food. Special puppy formulas are high in protein. Dry puppy food can also be used. One may want to mix/some of the puppy replacement milk formulas that are available with the dry puppy food or simply soak the food with water for an hour to soften it before feeding.

The amount to feed a puppy varies depending on its breed. Usually, a puppy should be fed three or four meals a day. A young puppy must receive enough food to provide its energy requirements and for adequate growth. Puppies experience their most rapid growth during the first 6 months of life. One must be careful not to over feed the puppy, because the puppy may become overly fat or grow too rapidly. Over feeding can lead to obesity and may worsen the symptoms of hip dysplasia and other inherited bone diseases. The bones of a young, overweight puppy may not be developed to adequately carry excess weight.

As a dog gets older, its metabolism slows down and its need for calories decreases. Owners must feed a ration containing adequate protein and nutrients to maintain body condition. One should guard against overfeeding, which will lead to obesity and cut down on the expected life of the animal. Older dogs should receive smaller meals, which are easier to digest, several times a day.

8. What are some of the general grooming and health care practices necessary to keep a dog clean and healthy?

The amount of time spent on grooming depends on the type of hair coat; daily brushing is recommended to remove dead hair and distribute the skin's oils. Brushing down to the skin will help remove flakes of dead skin and dandruff.

Long-haired dogs, in addition to regular grooming and brushing, need to be checked for mats of hair, which commonly occur behind the ears and under the legs. These masses of hair can usually be teased out with a comb but occasionally need to be cut out. Care should be taken when cutting out mats so that one does not cut the skin. Long-haired dogs also have problems with burrs from plants. When cutting the hair, one should slide the comb down under the mat of burr and cut the hair on the outside of the comb. This method will be less likely to cause injury to the dog.

The terriers and wire-haired breeds need their coats plucked to remove dead hair. Plucking is accomplished by using a stripping knife, grasping a section of hair between knife and thumb, and pulling the knife away with a twisting motion. This will remove dead hair and trim live hair.

Dogs should be bathed only when they become extremely dirty; frequent bathing will remove natural oils and cause the coat to become dry and harsh. When bathing becomes necessary, pH-balanced shampoo especially developed for dogs should be used. Detergent soaps should never be used because some dogs may have skin reactions to them. Many of the shampoos for pets are medicated and help prevent external parasites. When washing, one must be careful not to get shampoo in the dog's eyes. After bathing, it is important that the dog not become chilled. One should be careful of the temperature if bathing outside; if bathing inside, one should make sure the dog is completely dry before letting it outside.

The dog's nails will need to be trimmed occasionally. How often depends on the surface on which the dog is kept. Outside dogs usually will wear their nails down naturally on the soil surface; inside dogs need to be checked more often. Nails can be clipped at home using clippers obtained from a veterinarian or pet shop.

Clippers should be sharp so that they cut the nail and do not crush it; crushing the nail may cause pain to the nail bed. One must be extremely careful not to cut into the nail bed because this will cause bleeding and pain. With clear or white nails, the nail bed can easily be seen. Extreme caution needs to be taken when clipping black nails. Bleeding can be stopped using a styptic powder. One should check the dew claws on the inside of the leg; these will not wear down and many times are over looked.

Sharp, pointed scissors should never be used to trim nails because the dog may be injured if it should happen to move suddenly.

During routine grooming, one should check the ears; a dog's ears will need to be cleaned about once a month. A cotton swab or soft cloth soaked in mineral oil, hydrogen peroxide, or alcohol can be used. Only the part of the ear that can be seen should be cleaned. Never poke any sharp or hard object into the ear; a quick head movement could cause serious injury. Spaniels and other dogs with long hair need to be checked for mats and burrs that may block air movement and cause infections.

Irritating substances can be removed from the eyes with contact lens solutions or other eyewash solutions. Dogs with large, protruding eyes and long hair need special care to keep hair from rubbing the eyeballs. Hunting dogs and other field dogs should be checked carefully for irritating substances after each outing. Serious irritations and injuries need to be brought to the attention of the veterinarian.

A dog's teeth are usually not prone to decay. However, dogs do develop plaque and tartar that can lead to periodontal disease. Periodontal disease is painful and may lead to premature tooth loss. A dog's teeth should be cleaned once or twice a week. This can be accomplished by using a small toothbrush or gauze pad.

Toothpaste made for dogs, salt water, or an equal mixture of salt water and baking soda can be used; the teeth should be scrubbed from gum to crown. Regular cleaning will prevent many problems. If the plaque is not removed, it will develop into a hard yellow-brown or gray-white deposit on the teeth called tartar. Periodically, the dog will need to have its teeth professionally cleaned at the veterinarian's office to remove plaque and tartar both above and below the gum line.

Feeding dry dog food and some of the various treats on the market also helps prevent the buildup of tartar.

9. **What are the six groups of diseases that affect dogs?**
   - Infectious diseases
   - Noninfectious diseases
   - Fungal diseases
   - Internal parasites
   - External parasites
   - Poisonings

10. **List the diseases belonging to each of the six groups. What are the symptoms of each disease?**
    Infectious Diseases
    - Canine distemper
    - Infectious canine hepatitis
    - Leptospirosis
    - Canine parvovirus infection
    - Infectious tracheobronchitis
    - Rabies
    - Coronavirus
    - Canine brucellosis
    - Canine herpes virus
    - Salmonellosis
    - Haemobartonellosis
    - Gastrointestinal campylobacteriosis

    Noninfectious Diseases
    - Heart disease
    - Cataracts
    - Glaucoma
    - Progressive retinal atrophy (PRA)
    - Cherry eye
    - Hip dysplasia
    - Arthritis
    - Tetanus
    - Botulism
    - Anal sac blockage

    Fungal Diseases
    - Ringworm
    - Blastomycosis
    - Histoplasmosis
    - Coccidioidomycosis

    Internal Parasites
    - Ascarids, commonly called "roundworms"
    - Hookworms

- Whipworms
- Tapeworms
- Heartworms

External Parasites
- Fleas
- Ticks
- Lice
- Mites
- Chiggers

Poisonings

Symptoms of poisoning vary depending on the type of poison consumed. Treating animals that have been poisoned must begin as soon as a pet owner recognizes a poison is involved; prompt diagnosis and treatment are extremely important.

11. **What are some of the general care practices necessary for a pregnant female before, during, and after whelping? What are some practices for the general care of the new litter?**

A whelping box provides a nice place for the female to have her puppies. She should be introduced to her box about 3 weeks before whelping so that she can get comfortable with it. The bottom of the box should be covered with newspaper.

A normal, healthy female usually gives birth easily without need of additional help; however, one should stand by and give assistance if necessary. One must be prepared to help the mother remove a puppy from its placental membrane, to clean it, and encourage it to nurse. When cleaning a membrane off a puppy, the owner may also have to sever the umbilical cord. A piece of cotton thread should be tied around the cord about ¾ to 1 inch away from the puppy's body; the cord should then be cut ¼ to ½ inch beyond the thread, taking care not to pull on the cord and making sure the hands are clean. An application of iodine to the naval cord is recommended. One must be careful of the mother also because she may be protective of the litter and may bite.

The first milk produced by the mother is called colostrum. It is very important that each puppy receive colostrum because it contains immunoglobulins that will help protect it from infectious disease.

Eclampsia is a calcium deficiency that may be observed shortly after whelping and during the first month of lactation. It usually occurs among toy breeds and females with large litters that require a large milk production. Symptoms include restlessness, panting, whining, and getting up and down. As the problem worsens, the dog will become still, experiencing twitching muscles, and lack coordination. If not treated, she will collapse into convulsions and die. A veterinarian must be consulted immediately to give her a calcium injection. Prevention can be accomplished by ensuring the pregnant female gets proper vitamin and mineral supplements, especially calcium.

CHAPTER 8

Cats

DISCUSSION QUESTIONS AND ANSWERS

1. **What is the history of the cat?**
   Cats are descendants of the same carnivorous, tree-climbing mammal *Miacis* from which dogs descended. Cats, like dogs, trace their lineage through *Cynodictis*. *Proailueus,* the first normal or feline cat, appeared about 35 million years ago. The exact evolution of the cat from that time forward is not known, but modern cat types, such as *Felis* and related genus groups with "normal" canine teeth, appeared about 7 million years ago.

   It is believed that domestication of the cat began around 4,000 years ago, although there is no clear documentation until around 2500 BC. Early Egyptian paintings and inscriptions suggest that the cat was kept in captivity, tamed, and eventually revered and protected. The cat was used in religious ceremonies, and many mummified cats have been found in Egyptian tombs.

   As civilization spread from Egypt and the Middle East, domestication of the cat spread also. The Romans introduced the cat into Europe, and European explorers, travelers, and traders carried the cat to all parts of the world. Today, the domesticated cat can be found on every continent except Antarctica.

   It is believed that the domesticated cat of today originated from the African Wild Cat (*Felis libyca*). This cat was common to all areas of Asia and North Africa. Two other cat species were the Jungle Cat (*Felis chaus*) and the European Wild Cat (*Felis silvestis*).

2. **What are the two major groups of domesticated cats?**
   Short-haired breeds and long-haired breeds.

3. **List the different breeds of cats. How do they differ?**

   Short-haired breeds
   - Abyssinian
   - American Shorthair
   - American Wirehair
   - Bombay
   - British Shorthair
   - Burmese
   - Chartreux
   - Colorpoint
   - Cornish Rex
   - Devon Rex
   - Egyptian Mau
   - Exotic Shorthair
   - Havana Brown
   - Japanese Bobtail
   - Korat
   - Malayan
   - Manx
   - Ocicat
   - Oriental Shorthair
   - Russian Blue
   - Scottish Fold
   - Siamese
   - Singapura
   - Snowshoe
   - Sphynx
   - Tonkinese

   Long-haired breeds
   - Balinese and Javanese
   - Birman
   - Cymric
   - Himalayan and Kashmir
   - Maine Coon
   - Norwegian Forest Cat
   - Persian
   - Ragdoll
   - Somali
   - Turkish Angora

The cat breeds differ in: length of hair, color and color patterns, shape and length of ears, shape and color of eyes, shape of the head, conformation of the body, size, and disposition.

There is a miscellaneous class that includes the American Curl, the Turkish Van, the LaPerm, the Selkirk Rex, and the Siberian.

For specific differences between the breeds, consult the text.

4. **What is meant by the term *colorpoint*?**
Colorpoint refers to the color of the extremities: nose, ears, feet, and tail. Colorpoint cats have one general body color, and the extremities are a different, darker color.

5. **How should the long-haired cats be groomed? Do short-haired cats need grooming?**
Long-haired cats should receive daily care. If neglected, the hair will tangle and mat. Removal of these tangles and mats may be difficult and an unpleasant experience for the cat.

Equipment for long-haired cats should include a comb with two sizes of teeth, a fine-tooth or flea comb, nail clippers, a grooming brush made with natural bristles (nylon may cause excessive static), and grooming powder (baby powder, talcum powder, or cornstarch).

Using a wide-toothed comb, comb all areas of the animal. One must be careful of the sensitive areas of the stomach, the insides of the legs, and under the tail. If the coat is free of tangles and mats, the fine-toothed part of the comb should be used. The skin and not just the outer fur should be combed; next, the fur should be brushed out. One should brush in the opposite direction to which the hair naturally lies and occasionally sprinkle grooming powder into the fur.

If the fur has become badly tangled and matted, scissors should be used to cut the mats out, being careful not to injure the animal. Blunt-ended scissors—never sharp-pointed scissors, knives, or razor blades—should be used. Do not cut parallel to the skin. Split mats into strips by cutting perpendicularly away from the skin. A sudden movement by the cat could cause serious injury. Badly matted cats may require sedation and clipping by a vet.

Grooming short-haired cats can usually be accomplished with a fine-toothed or flea comb. In many cases, hand grooming will be sufficient to remove dead hair. A rubber grooming brush is also very effective, but it must be used carefully because good hair may also be removed. The use of a soft chamois, silk, or nylon pad causes some static in the coat and helps it cling tightly to the body.

6. **What is the most important thing to remember in feeding a cat?**
The most important thing to remember in feeding cats is that they require almost twice as much protein in their diets as do dogs. The best source of this is from animal products; 30 to 40 percent of the cat's diet should be animal-type proteins (meat, meat by-products, fish, eggs, and milk).

7. **What are the six groups of diseases that affect cats?**
   - Infectious diseases
   - Noninfectious diseases
   - Internal parasites
   - External parasites
   - Fungal diseases
   - Poisonings

8. **List the diseases in the six groups that affect cats. What are the symptoms of these diseases?**
   - Feline panleukopenia
   - Feline herpes virus and feline calicivirus
   - Feline rhinotracheitis
   - Feline infectious peritonitis
   - Feline leukemia virus (FeLV)
   - Feline pneumonitis

- Rabies
- Feline urologic syndrome
- Wet eyes
- Toxoplasmosis
- Ascarids and hookworms
- Tapeworms
- Lice (*Felicola subrostratus*)
- Mites (*Demodes cati*)
- Feline scabies or notoedric mange (caused by *Notoedres cati*)

9. List the general health care practices for a pregnant cat before, during, and at birth.

About 2 weeks before the kittens are due, the female becomes restless and searches for a quiet place to give birth. If one offers her a well-prepared box, she will usually take to it.

Normally, very little assistance is needed for the mother during delivery. Occasionally, a female does not attempt to remove the sac from the newborn kitten. She may not know what to do or may be too busy with the next kitten. In this case, one must remove the membrane. If the female still does not show interest, one should carefully sever the umbilical cord, rub the kitten dry, and stimulate it to breathe. Then, the kitten should be placed close to the mother's nipples. Hopefully, the kitten will begin to nurse and arouse maternal instincts in the female.

10. What health care practices should be observed with the new litter?

The first milk produced by the mother contains colostrum. Kittens receive temporary immunity from many diseases from the antibody-rich colostrum in their mother's first milk.

CHAPTER 9

Rabbits

DISCUSSION QUESTIONS AND ANSWERS

1. **What is the history of the rabbit?**
   The European wild rabbit is the species from which all domestic rabbit breeds have been developed. Rabbits were fairly widespread and abundant in Europe during the late Tertiary period and early Pleistocene epoch. Abnormal climates during the Ice Age drove them to the southern parts of Europe.
   First reports of rabbits were from Phoenician traders who visited the coast of Spain and the island along the coast around 1100 BC. The Phoenicians are probably responsible for transporting rabbits to other parts of the world.
   Rabbits were of great economic importance. They were hunted for food, and their pelts were used to make clothing. Credit for domestication of the rabbit is given to French monks of the Middle Ages, who raised rabbits in walled cages kept in their monasteries. They served as an easily raised source of food, and the fur was used for clothing.
   In the second half of the nineteenth century, the European wild rabbit was introduced into Australia and New Zealand, where it quickly spread and became a serious pest. Brought to Chile in the early twentieth century, it eventually spread over much of the South American continent. Some European wild rabbits were released on the San Juan Islands off Washington State in the United States at the beginning of the twentieth century and have flourished since, but the major rabbit of North America remains the cottontail.

2. **What are the uses for rabbit?**
   Rabbits are used for meat, research, fur or wool, and pets.

3. **What are the advantages of rabbit meat over other types of meat?**
   Those who encourage the use of rabbit meat stress the following advantages: high in protein; low in cholesterol, fat, and sodium; and very palatable.

4. **What are the five weight categories of rabbits?**
   The five weight categories are dwarf, or miniature; small; medium; large; and giant.

5. **List the breeds in each of the weight categories.**

   Dwarf or Miniature
     Brittania Petite
     Netherland Dwarf
     Himalayan
     Dwarf Hotot
     Polish
     Jersey Wooly
     Holland Lop
     American Fuzzy Lop
     Mini Rex

   Small
     Dutch
     Tan
     Florida White
     Silver
     Avana
     Mini Lop

Medium
    English Spots
    Standard Chinchilla
    English Angora
    Lilac
    Silver Martin
    Belgian Hare
    Rhinelander
    Harlequin
    Sable
    Satin Angora
    French Angora
    Rex

Large
    Beveren
    Californian
    Hotot
    Palomino
    Satin
    Cinnamon
    Creme d'Argent
    Champagne d'Argent
    American
    American Chinchilla
    English Lops
    New Zealand
    Silver Fox

Giant
    Giant Angora
    French Lops
    Checkered Giant
    Giant Chinchilla
    Flemish Giant

6. **What are the houses for rabbits commonly called?**
Houses used to keep rabbits are referred to as *hutches*.

7. **What are several important things to remember in housing rabbits?**
Several things are important in housing rabbits: temperature, humidity, ventilation, proper lighting, and absence of drafts.

8. **Describe how a rabbit should be handled.**
The correct way to handle a rabbit depends on the size of the rabbit; however, under no circumstances should the animal be picked up by the ears because this may cause injury to small blood vessels or cartilage in the ear. Also, picking the animal up by the nape of the neck is not recommended, because this too can cause damage to tiny blood vessels in the skin and could cause damage to the pelt.

    When approaching a rabbit cage, one should walk slowly and speak to the rabbit. If the rabbit knows someone is close, it won't get as frightened and bolt to the corner of the cage.

    Next, one should reach in and gently stroke the animal from front to rear and rub and stroke the animal's head. Then, the rabbit can be gently moved into a position where it can be picked up by slipping one hand in under the chest and belly and placing the other hand behind the rabbit. The rabbit can now be lifted up and removed from the cage. It is important to remove it from the cage tail first to prevent the rabbit's legs from getting caught in the wire floor or wire door of the cage.

    Small rabbits can be supported with one hand under the body, and large rabbits can be supported on one arm with the other arm cradling the animal securely. If the animal feels secure and comfortable, it won't struggle or try to escape. The head of the rabbit can be tucked in under the upper arm toward the elbow of the arm that is being used to support the rabbit.

    Correct restraint is important because a frightened rabbit can kick hard enough to fracture its own spine.

    Small animals are easy to lift with the method described; however, larger animals may pose a problem. This is especially true if they are not used to being handled. The handler should always wear long-sleeved clothing to prevent scratches.

    When setting the rabbit down, one must do so gently and slowly, letting the animal see where it is going so that it will not be frightened. The handler should set the rabbit down, hind end first. A rabbit's foot pads are covered with fur, and when placed on slick surfaces, they have no traction. Rabbits can easily dislocate a hip or their spine when they try to hop or push off on their rear legs.

The handler should be careful when returning the rabbit to its cage. The rabbit, recognizing a familiar area, may try to jump from the handler's arms and may inflict an injury to the handler with its strong, powerful rear legs and sharp toenails.

9. **List and describe the common diseases and ailments of rabbits.**
   - Enteritis
   - Snuffles
   - Mastitis
   - Weepy eye
   - Treponematosis (vent disease, rabbit syphilis)
   - Ear mites
   - Infectious myxomatosis
   - Papillomatosis
   - Ringworm
   - Fur mites
   - Mange mites
   - Pinworms
   - Wet dewlaps
   - Fur chewing
   - Hutch burn
   - Sore hocks
   - Malocclusion

CHAPTER 10

Hamsters

DISCUSSION QUESTIONS AND ANSWERS

1. **What is the history of the common hamster?**
   Golden hamsters are native to the desert areas of Syria and are sometimes referred to as Syrian Golden hamsters. They were discovered in 1930 by Professor I. Aharoni of the Department of Zoology, Hebrew University, in Jerusalem. Professor Aharoni was exploring an animal tunnel near Aleppo, Syria, when he came across a mother and her litter. Of the litter of 12, only a male and two females survived the trip back to the university. These three were later mated and are believed to be the foundation for all Golden hamsters in captivity today.
   In 1931, young hamsters were shipped to the United States Public Health Service Research Center at Carville, Louisiana, where they were used in medical research. While being used in research, it was discovered that the Golden hamster could be tamed and made into a pet.

2. **What are the common types of hamsters found in pet shops?**
   The Golden hamster, *Mesocricetus auratus,* is the most abundant of the hamsters. This species has been most often used in research and is the species most commonly found in pet stores.
   Another species of hamster that can be found in pet shops is the small desert or dwarf hamster, *Phodopus sungorus.* The dwarf hamster is light gray with a dark stripe down the back. It is smaller than the Golden hamster.

3. **What foods can be fed to hamsters?**
   The easiest way to feed hamsters is to purchase commercially prepared pellets specifically designed for small animals.
   In mixing a ration, it is important that the animal receive a wide range of foods to ensure a nutritionally balanced ration. The diet should contain seeds such as corn, millet, wheat, oats, sorghum, and rapeseeds. Carrots, pieces of potato, fresh clover, or alfalfa hay should be supplemented into the diet as well as dried peas, beans, and nuts. Dry dog pellets or biscuits can be fed also.

4. **How should a hamster be handled?**
   Hamsters are nocturnal animals and should not be disturbed while sleeping. If a sleeping hamster is poked or touched, it may awaken quickly. Its reaction is to bite in self-defense. Occasionally, a hamster may need to be awakened during the day to clean its cage. To awaken a sleeping hamster, one should approach it carefully, tap lightly on the cage, and speak to it. A hamster learns to recognize its handler's voice, and speaking to it will calm it. The handler should attempt to stroke or pet it and see how it reacts. It may allow itself to be picked up; however, if it retreats to a corner or stands on its rear legs with its paws upraised, one must be very careful because this is the hamster's defensive posture. The handler should continue to talk to the animal and, if it leaves the corner, try to pet or stroke it. When the hamster allows it, it can be picked up with the thumb and forefinger placed right behind the front quarters. Hamsters are quite strong and very dexterous in their attempts to escape. They can twist their bodies and use their rear feet to free themselves. The more a hamster is handled, the tamer it will become.

5. **List the common diseases and ailments of hamsters. What are the symptoms? How can we control each of them?**
   - Wet tail
   - Tysser's disease
   - Common diarrhea
   - Fleas and lice
   - Mites
   - Lymphocytic choriomaningitis (LCM)

6. **What common mating method is used with hamsters?**
   Two methods are commonly used when mating hamsters: individual mating and colony mating. In individual mating of hamsters, the female should always be placed in the male's cage; this will usually result in less fighting. After mating, the male will rest for a few moments, clean himself, and then pursue the female again. They may mate several times in a 15- to 20-minute period. When the female is satisfied, she will turn on the male and bite him. At this time, the female should be removed from the male's cage and placed back in her cage.

CHAPTER 11

Gerbils

DISCUSSION QUESTIONS AND ANSWERS

1. **What are the general characteristics of gerbils?**
The Mongolian gerbil will reach a size of 6 to 8 inches long from nose to tail; the body is about 3 to 4 inches long. A mature gerbil will weigh 3 to 4 ounces.

The body is short and thick, and the gerbil has a hunched appearance when squatting on its hind legs. The forelegs are very short, with the forepaws being very similar to hands and used to hold food. The hind legs and feet look very similar to those of a kangaroo and are large and furry with long toes and claws. The large feet enable the gerbil to stand firmly in the sand of its desert environment. The gerbil walks on all fours but, when stopped, will stand up on its hind legs to observe the immediate area. In the wild, gerbils use their strong hind legs to leap and jump, but those raised in captivity will rarely jump, except in efforts to escape their cages.

The tail is covered with fur, has a bushy tip, and is used for support when standing; it also acts as a rudder and stabilizer when jumping or leaping. The skin covering the tail is loose and, if handled roughly, can become torn and pulled from the tailbone.

The head is broad and wide, and the eyes are large, usually jet black, and protrude slightly. The ears are small, rounded, and covered with soft fur. The gerbil's sense of hearing is very acute. This is because of the enlarged cavities in the skull that amplify the slightest sounds.

The natural color of the Mongolian gerbil is a reddish brown to dark brown. The fur is composed of dark guard hairs that give the overall coat a darker appearance. The underside fur is usually lighter brown, cream, or even white. This natural color pattern is referred to as an agouti color and serves as a camouflage against the sand and rock of its environment. The light underside fur serves to reflect heat from the desert sand.

Through mutations and selective breeding, gerbils are found today in several different color patterns. Black, white, black and white, white-spotted agouti, charcoal and white, and orange and white are common color patterns. The red-eyed, white gerbil is not considered a true albino because some of the fur on the back and tail will turn light brown as the animal matures.

Both male and female gerbils have a scent gland on their stomachs that looks like a bald patch or strip; this patch is more visible on mature males. The adults leave their scent by sliding their stomachs along an object or area.

Gerbils are very quiet animals. A shrill squeak is sometimes made as an alarm and during mating by the female. Young gerbils can be heard making squeaking sounds, but these sounds will diminish as they get older.

Another method of communication is a drumming sound made by standing upright and pounding with the rear feet; this drumming sound is also used as an alarm and by the male during mating.

Gerbils are short-lived animals. The life expectancy is usually 2 years but may be up to 4 years.

2. **How do hamsters and gerbils differ?**
   There are two main differences between hamsters and gerbils. Gerbils have long tails, and hamsters have very short tails. Gerbils have large rear feet on which they can easily stand in an upright position.

3. **What types of food are usually fed to gerbils?**
   The easiest way to feed gerbils is using commercial pellets made especially for small rodents. These will be complete balanced rations, with all the required vitamins and minerals added.

   If preferred, rations can be prepared by mixing different types of foodstuffs. Common grains used are corn, oats, wheat, and barley. Mixing these grains can be a good base for a balanced diet. Gerbils will eat only the heart out of the corn kernel, and just the kernel from oat seeds. Waste can be prevented by feeding breakfast foods made from corn flakes and rolled oats.

   Linseed, millet, canary seed, rapeseed, and hemp seed found in mixtures for parakeets and other birds add variety to the diet.

   Green foods such as cabbage, carrots, turnips, and beets also add variety to the diet and are usually rich in minerals.

4. **How should one pick up and handle a gerbil?**
   The easiest way to pickup a gerbil is to place your hand over the gerbil's back and encircle his body with your thumb and forefinger. While holding the gerbil, one should stroke the head and back. The animal should be calm before any attempt is made to pick it up. One should be careful not to squeeze the animal too tightly, because it will struggle to escape. Many times, a gerbil will nibble on a handler's finger, not to inflict a bite but to let the handler know it has been held long enough and wants down. After a gerbil has become accustomed to being handled, it will normally climb right into the owner's hand.

   A gerbil should never be picked up the tail, because the skin can tear and pull loose, exposing the tailbone. The exposed portion of the tail will normally die and fall off.

5. **What are the common diseases and ailments of gerbils? What are their symptoms and how should they be treated?**
   - Colds
   - Red nose
   - Tyzzer's disease
   - Parasites
   - Fits or seizures

CHAPTER 12

Rats

DISCUSSION QUESTIONS AND ANSWERS

1. **What are the two species of rat that have been domesticated?**
   The black rat—sometimes referred to as the roof, climbing, or gray rat—and the brown rat—sometimes referred to as the Norway, barn, sewer, or wharf rat.

2. **Describe the common types of rats used for research and as pets.**
   Because of mutations and selective breeding, both the black and brown rat can be found in several colors. The white laboratory rat is a descendant of albino strains. Creams, fawns, and light-gray varieties are common today; another common variety is the hooded rat. It has a colored head and a dorsal stripe on a white body; the colors are brown, black, fawn, and cream. The caped variety has a colored head but lacks the colored dorsal stripe.

3. **How should one handle pet rats?**
   Young rats can easily be picked up by grasping them around the body just behind the front legs. They may also be picked up by grasping the tail at the base close to the body and lifting.
      When handling older rats, be careful and talk so that they will relax. How the handler approaches the animal depends a great deal on how tame the animal is. If the animal is not tame, one must go slowly, get its attention, grasp it by the base of the tail, and lift. If further restraint is needed, the handler should take the other hand and grasp the animal around the body just behind the front legs and restrict the movement of its head with the thumb and forefinger.
      A rat should never be grasped by the tip of the tail. In an effort to free itself, it will twist and squirm. Its tail or the skin on the tail can break loose and be pulled off, leaving the tailbone exposed. The rat is very dexterous and can turn and climb up its tail. Leather gloves should be used if unsure of the situation.

4. **What are the common foods for pet rats?**
   The easiest way to feed rats is to use commercially prepared pellets formulated for small animals; these contain all of the nutrients that are necessary. Commercial rodent diets can be supplemented with small amounts of dry dog food, crackers, cereal, fruits, or vegetables. Overfeeding a ration of fruits and vegetables may cause diarrhea. In general, such treats should compose less than 10 percent of the daily diet.

5. **What are the common diseases and ailments of pet rats?**
   - Respiratory disease
   - Parasites

CHAPTER 13
Mice

DISCUSSION QUESTIONS AND ANSWERS

1. **What is the history of the domesticated mouse?**
   The common mouse is the most familiar and most widely distributed rodent in the world; there are four subgenera and 36 species. The best-known species is the common house mouse (*M. musculus*); this species lives wherever humans live.

   The word *mouse* comes from an old Sanskrit word meaning "thief"; Sanskrit is an ancient language of Asia, where scientists believe the house mouse originated. The house mouse was originally a wild field animal. Grain and other foodstuffs stored in human settlements offered an attractive source of food for mice, which adapt quickly to new environments.

   From Asia, the house mouse spread throughout the world. The common house mouse is one of the world's most successful mammals; they multiply extremely quickly and are able to adapt to almost any condition. Their reputation has been as a dirty animal associated with carrying diseases and as a very destructive animal capable of causing tremendous economic losses of stored grains and other property.

   Subsequently, mice have gained a place in history. The ancient Egyptians and Romans had a phrase that described the mouse's reproductive ability; they used to say, "it's raining mice," or, "mice are made of raindrops." The Greeks and the people of India believed that mice were lightning bolts born of thunderstorms. In some places, mice were kept in temples and worshiped as sacred animals. They were viewed as instruments of punishment. People would present sacrificial offerings to stop or prevent a mouse plague.

   More than 4,000 years ago, the Cretans built a temple in Tenedos, Pontus, where they fed and worshiped mice. In the Cretan victory over Pontus, according to legend, mice helped by chewing through the leather straps of the shields of the Pontic soldiers so that they were unable to defend themselves.

   Paintings on ancient bowls and other clay artifacts tell us that Egyptians also kept mice about 4,000 years ago. The first white and spotted mice appeared about 4,000 years ago in Greece, Egypt, and China. White mice in particular were considered sacred; people used them to predict the future, as lucky charms, or to keep wild mice away. Waltzing mice originated in China 2,000 to 3,000 years ago. The Japanese bred white and colored mice systematically 3,000 years ago. In the 1800s, people in Europe began to exchange "fancy mice." Fancy mice have different coat colors and patterns.

   In ancient Rome and during the Middle Ages, mice were used as medicine against all kinds of diseases. They were dried, pulverized, sliced, or marinated in oils, and then used as a compress or taken internally. Their blood was a favored ingredient of drugs and tonics. Mice were supposedly good for flesh wounds, snake bites, warts, bladder irregularities, diabetes, enlarged thyroid glands, diseases of the eye, and loss of hair.

   Even though mice were used for medicinal purposes, they were also targets for extermination. All kinds of methods were recommended to get rid of them, except for the cat. During the Middle Ages, cats were not recognized as natural predators of the mouse nor were they considered pets. Cats were hated as accomplices of sorcerers and other witches and suffered extreme cruelty.

In more modern times, the house mouse, especially the albino strains of *M. musculus,* has been widely used for medical and biological research, particularly in the study of heredity.

By careful breeding over the years, today's mice are more gentle and less timid than their ancestors. Pet mice are relatively free of diseases, and if handled frequently, they show little tendency to bite or escape. Today, these curious and interesting mammals are available in many different colors and color combinations.

2. **Mice were used by ancient Rome for medicinal purposes. What were they used to treat?**
In ancient Rome and during the Middle Ages, mice were used as medicine against all kinds of diseases. They were dried, pulverized, sliced, or marinated in oils, and then used as a compress or taken internally. Their blood was a favored ingredient of drugs and tonics. Mice were supposedly good for flesh wounds, snake bites, warts, bladder irregularities, diabetes, enlarged thyroid glands, diseases of the eye, and loss of hair.

3. **What are the characteristics of the common house mouse?**
The mouse has a pointed nose and a split upper lip. Mice have four toes on their forefeet and five toes on their hindfeet.

The house mouse is approximately 2½ to 3½ inches long excluding the tail; the tail is generally the same length as the body. This small mammal will weigh ½ to 1 ounce. The house mouse has a small head and a long, narrow snout. Several long, thin whiskers are used by the animal to help find its way about in dark, tight areas. The fur of the house mouse is grayish-brown on the animal's back and lighter colored to white on the underside.

The eyes of the house mouse are fairly large, round, and black; they cannot see very well. Because of the large, spherical shape of the eye, they have an almost prefect visual field, but because of the position of the eyes on the side of the head and the shape of the eye, they are incapable of detailed vision. Mice can detect movement but may be unable to determine what the movement is. Cats, in their pursuit of a mouse, move ever so slowly and may remain motionless for long periods.

Mice have fairly large ears and possess a highly developed sense of hearing. They perceive sounds in very high frequency ranges, up to 100,000 Hz.

Smell is the most highly developed sense that mice possess. Mice use smell for locating food, identifying family and colony members, finding their way around in darkness, and identifying enemies.

Mice are primarily nocturnal in habit and will usually seek hiding places during the day, but pet mice, feeling safe around humans, may come out during the day. Pet mice can recognize by smell those who feed and handle them and will feel at ease in their presence.

4. **What is a clan or colony structure? Describe the advantages of such a structure.**
Mice are gregarious and prefer the company of other mice. The house mouse lives in groups that are called clans or colony structures.

This colony structure gives the colony more protection against enemies, and it is easier for the colony members to find and store large quantities of food. Females in the colony will share duties of raising the young, providing warmth for the young, and nursing and feeding the young.

5. **What is the main diet of mice in the wild?**
Mice in the wild will eat almost anything, but their main diet consists of grain and seeds.

6. **How should a pet mouse be handled?**
A mouse should not be picked up by the tail or nape of the neck because this causes discomfort and pain. If necessary, a mouse can be grasped by the base of the tail and lifted. Another method of getting a mouse out of a cage until it gets used to being handled is to scoop it up in a cup.

A mouse will usually go into a cup because of its curious nature.

When purchased at a young age, a mouse will tame very quickly; it will take food from fingers and eventually climb onto the owner's hand.

7. **What are the common foods for pet mice?**

   Commercially prepared pellets or a mixture of grain and seeds can be used. Commercial feeds come in a hard pellet form and contain all necessary nutrients. The hard pellet is also beneficial in keeping the continuously growing front incisors worn down.

   A ration can be purchased or prepared using corn, oats, and wheat and then adding small amounts of millet, barley, and buckwheat. Small amounts of oily seeds are good for adding shine to the hair coat. Commercially prepared feeds are recommended because the amounts and ingredients are chosen to ensure that the diet is complete and balanced.

   In feeding a ration, various greens and vegetables can be added. Fresh grass, lettuce, dandelions, carrots, apples, dates, and raisins can be added. Clean, fresh hay can supplement the diet and will also serve as a nesting material.

8. **What are the common diseases and ailments of pet mice? What are the symptoms and treatments for them?**
   - Respiratory diseases
   - *Salmonella*
   - Parasites

CHAPTER 14

Guinea Pigs

DISCUSSION QUESTIONS AND ANSWERS

1. **Where were guinea pigs originally found?**
   The domesticated guinea pig has been bred for meat production in South America for at least 3,000 years. The range of domestication extended from Northwestern Venezuela to Central Chili. The guinea pig is still widely kept as a source of food by the native people of Ecuador, Peru, and Bolivia.

2. **Where does the name *guinea pig* come from?**
   The guinea pig is neither a pig nor does it come from Guinea; the actual reason for the name is not known, but there are several possible explanations. The species name *porcellus* (*Cavia porcellus*) for the domestic guinea pig means "little pig" in Latin. The only similarities to a pig are the low grunts, the squeals they make when they are hungry, and their fat little bellies.

    After obtaining some of the animals, Dutch and English traders stopped off in Guinea on their way back to Europe. The first guinea pigs sold in England for one English coin called a *guinea*. Another explanation could be that guinea comes from the Portugese word *guine*, which means "far away and unknown lands."

3. **What is the more correct name for the guinea pig?**
   The guinea pig is more accurately called a cavy.

4. **What are the seven varieties of guinea pig? What are the characteristics of each type?**
   The seven common varieties of guinea pigs are the Abyssinian, American, Peruvian, Satin, Silkie, Teddy, and White Crested.
   The Abyssinian has a rough, wiry hair coat. The hair is made up of swirls or cowlicks called rosettes. The more rosettes on the animal, the more desirable. The Abyssinian is found in all colors and color combinations.
   The American is the most common of the guinea pig varieties. The hair of the American is short, very glossy, and fine in texture. This short-haired guinea pig can be found in a wide variety of colors and color combinations; its short hair makes it easy to care for.
   The Peruvian is the long-haired variety. Length, evenness, and balance of the hair coat are deciding features when judging this variety. The Peruvian can be found in many of the same colors and color combinations as the short-haired, American variety. The hair coat takes a great deal of time and effort to keep clean and may reach a length of 20 inches. Because the guinea pig does not have a tail, the Peruvian has the appearance of an animated mop, and it may be difficult to tell which end is which.
   The luxurious, shiny fur of the Satin is its major characteristic. The coat is fine, dense, soft, and has a sheen. Satins are found in the same colors and color combinations as the other guinea pig varieties.
   The Silkie is another long-haired variety, but unlike the Peruvian, there is no long, frontal sweep over the head. The Silkie has a mane that sweeps back from the head, between the ears, and back over the back and down the sides. The Silkie is also found in the same colors and color combinations as the other varieties.

The Teddy variety has short, kinky hair. The fur is short, resilient, and thick and lies close to the body; the whiskers are also kinked. The Teddy can be found in the same colors and color combinations as the other guinea pig varieties.

The White Crested guinea pig has short hair and resembles the American short-hair guinea pig except for the crest, which is a rosette and radiates evenly with a clearly defined center from the forehead. The White Crested is primarily found in self, solid, and agouti colors; the crest is, of course, white.

5. How should a guinea pig be picked up?
A guinea pig should be grasped firmly around the front shoulders with one hand and have its rear supported with the other hand. This will prevent the guinea pig from struggling. Guinea pigs can injure themselves if allowed to struggle while being picked up with one hand. Once picked up, it should be cradled in the palm and forearm and held close to the body. In this position, the animal will feel safe and secure. The handler must not allow the animal to fall; guinea pigs are not agile and can injure themselves if allowed to fall on a hard surface.

6. What is the importance of vitamin C in the guinea pig diet?
Guinea pigs, unlike many other animals, cannot synthesize vitamin C in their bodies and must be supplied with it in their diets. A lack of vitamin C will cause scurvy, which results in dehydration, poor appetite, diarrhea, rough hair coats, lethargy, and weight loss. Small, pinpoint hemorrhages on the gums and joint swelling with lameness also occur. The animals have severely reduced resistance to infectious agents causing respiratory disease. Most commercially prepared pellets will have vitamin C added; however, it breaks down rapidly, and opened bags of pellets should be used within 30 days.

7. What are the common foodstuffs for guinea pigs?
Guinea pig pellets are available that contain all the nutrients needed. These pellets are hard and serve to wear the teeth down.

Pellets that have high alfalfa content and dry alfalfa hay are suitable for young, growing guinea pigs and pregnant females. Clover and grass hay pellets and dry clover and grass hays with low protein content are more desirable for grown animals. It is important that hay not be moldy, musty, or dusty, moldy hay can be fatal. The pellets and dry hay also help to wear the animal's teeth down.

Romaine lettuce, green leaf lettuce, carrots, apples, pears, turnips, beets, and cucumbers are also favorite foods of guinea pigs. Iceberg lettuce should be avoided, because it is less nutritious and may lead to diarrhea. Wheat, corn, and oats are high-protein foods that can be added to the diet.

8. What are the common diseases and ailments of guinea pigs? What are the causes, symptoms, treatments, and control methods?
- Respiratory diseases
- Toxemia
- Parasites
- Malocclusion

CHAPTER 15

Chinchillas

DISCUSSION QUESTIONS AND ANSWERS

1. **Where did chinchillas originate?**
   The native habitat of the chinchilla is the barren areas of the Andes Mountains at elevations up to 20,000 feet.

2. **What are the two species of chinchilla?**
   Chinchillas are classified into two species: *C. brevicaudata* and *C. laniger*.

3. **Who is responsible for the chinchilla industry in the United States?**
   Mathias F. Chapman, a mining engineer with Anaconda Copper Company, is the person responsible for introducing the chinchilla industry in the United States.

4. **What are the mutant colors of chinchilla?**
   The first important mutant was a white male born in 1955. Another mutant was a beige female born in 1955 on an Oregon ranch. Black mutations called Black Velvet were developed in 1956 by an American breeder. Other mutations are the Blue Black Velvets, Sapphire Velvets, Pastel Velvets, Brown Velvets, and Sullivan Violets. Common mutant colors include white, beige, and black.

5. **What are common foods for chinchillas?**
   Various pelleted foods are available today for chinchillas. These pellets consist primarily of ground alfalfa hay, corn gluten, corn tailings, wheat germ, and bran. The pelleted foods may be supplemented with fresh, well-dried alfalfa or timothy hay.

   The diet can also be supplemented with grass, carrots, and celery. Pears and apple slices will be readily consumed by chinchillas. As special treats, chinchillas love raisins.

6. **Describe a polygamous mating system.**
   Polygamous breeding cages are available for owners who want to breed more than one pair of animals. This type will usually have three separate cages connected together with holes in the back of the cages that lead to a tunnel that connects all three cages. Females are fitted with a plastic collar that prevents them from leaving their cages and entering the tunnel. Two females and one male are placed in the cages, and the male can use the tunnel to go from one cage to the other. If one of the females allows him to stay in her cage, a third female can be added to the cage where the male was first put. Additional cages can be added to house up to 20 females. Allowing the male to run the tunnel keeps him in good condition.

   In polygamous matings, the male is left in the tunnel and can visit the cages of the females at any time. The openings to the tunnel can be closed off to prevent a female from being rebred right after giving birth.

7. **List the common diseases and ailments of chinchillas. What are the causes, symptoms, and treatments?**
   - *Pseudomonas aeruginosa*
   - Conjunctivitis
   - Otitis
   - Pneumonia

- Digestive disorders
- Reproductive disorders
- Pathogenic organisms
- Parasites
- Impaction
- Fur chewing
- Malocclusionis
- Thiamine (vitamin B) deficiency
- Calcium to phosphorous imbalance or phosphorous deficiency
- Mastitis

CHAPTER 16

Ferrets

DISCUSSION QUESTIONS AND ANSWERS

1. **What is the ancestor of the domestic ferret?**
   The domestic ferret (*Mustela pulorius furo*) is believed to be a descendant of the European polecat (*Mustela pulorius*).

2. **What are the characteristics of the ferret?**
   The domestic ferret, like the other members of the genus *Mustela*, has a body that is elongated, lean, slender, and muscular. The legs are short, and the feet have five toes with claws.

   The head is oval-shaped, and the snout area is pointed. The head of the male usually is broader and less pointed than the females.

   Male ferrets (hobs) will be 16 to 20 inches long and weigh 3 to 5 pounds. Females (jills) are a little smaller, usually 12 to 14 inches long and weighing 1½ to 3 pounds.

   Ferrets do not have a very well-developed sense of sight; being primarily nocturnal in nature, they do not see well in bright light. They do, however, have highly developed senses of hearing, smell, and touch. When put in new surroundings, they will use their sense of smell to check out the area thoroughly. A ferret learns its owner's voice and will usually come to the door of a cage when the owner approaches.

   Ferrets that are well cared for will live from 8 to 11 years. Normal body temperature is 101.8°F; the heart rate is between 180 and 250 beats per minute, and respiration is between 30 and 40 breaths per minute.

   Ferrets have a total of 34 teeth. On each side, there are three incisors on top and three on the bottom, one canine tooth on the top and bottom, three premolars on top and three on the bottom, and one molar on top and two on the bottom.

3. **What are the eight color variations of ferrets?**
   There are a number of colors now being bred: sable (the most common), red-eyed white, silver mitt, sterling silver, white-footed sable, butterscotch, white-footed butterscotch, and the rare cinnamon.

4. **How should a ferret be handled?**
   When approaching a ferret, one should speak to it in a soft, gentle voice before attempting to pick it up. One should grasp the ferret firmly around the body behind the forelegs. The ferret will go limp and allow itself to be picked up. Place the other hand under the rear of the animal to support its body.

5. **What type of housing and equipment are necessary for a ferret?**
   The type of housing depends on the owner's situation. A single ferret should have a cage at least 12 inches wide, 24 inches long, and 10 inches high. This amount of space will sufficiently house a ferret and provide space for a litter, feeding, and sleeping area; however, a ferret is very active and, if kept in this size of cage, should be allowed out frequently for exercise.

   Cages that are 24 inches wide, 24 inches long, and 14 inches high are more desirable. A cage approximately 2 feet wide and 4 feet long is appropriate for a pair of ferrets; this allows plenty of room for a sleeping

or nesting box, a litter box, and space for exercising. Cages of this size can easily be constructed from wood and wire screen.

Glass aquariums are not desirable because they do not allow air to circulate and can become overly warm in the summer.

Whatever is used to house a ferret, two things should be kept in mind: (1) If there is a way to escape from a cage, a ferret will find it. All materials should be of good quality, and latches on the doors should lock securely. (2) Ferrets enjoy watching activity around them. Cages should be constructed to allow ferrets to see out and observe what's going on; they want to be part of the action.

Food bowls should be made of earthenware. These are heavy and almost impossible for the ferret to upset. A constant supply of fresh water should be available; the large vacuum-type, bottles used for rabbits and guinea pigs are ideal.

Ferrets are very playful, and almost anything will make a suitable toy. Toys that are available for cats will usually work well for ferrets, but ferrets are harder on the toys than cats. Ferrets chew on the toys and can break off small pieces. Then, they swallow those pieces, causing digestive tract blockages. Soft rubber toys should be avoided. Plastic pipes that they can crawl through and plastic milk jugs with holes cut in them provide ferrets with a place to explore and play.

6. **What are ferrets commonly fed?**
Ferrets are easy to feed. They are carnivores. Food for ferrets should be high in animal protein, relatively high in fat, and low in fiber. Appropriate ferret food should be about 35 percent protein, 20 percent or more in fat, and less than 3 percent fibers. Commercial diets are being formulated, but labels should be checked carefully. Dry foods are recommended, because they keep the ferret's teeth and gums in good condition, cost less than moist foods, and are easier to feed.

7. **What happens if a female ferret is not bred when she comes into heat? Why does this happen?**
When a female ferret comes in heat, she will remain in heat if she is not bred. The breeding cycle is controlled by induced ovulation, meaning the eggs are not released into the womb for fertilization until mating has taken place. The continued production of estrogen during the heat period can lead to aplastic anemia. The seriousness of this ailment is the females' loss of weight and the deterioration of her general condition. If the animal is not bred and she remains in heat, she could die. One should consult one's veterinarian about the proper course of action, which could be breeding, spaying, or bringing out of heat artificially.

8. **List the common diseases and ailments of ferrets. What are the causes, symptoms, and treatments?**
   - Canine distemper
   - Rabies
   - Digestive bacteria infections
   - Botulism
   - Parasites
   - Injuries

CHAPTER 17

Amphibians

DISCUSSION QUESTIONS AND ANSWERS

1. **What is an amphibian?**
Amphibians are that group of vertebrate animals that live the first or larval part of their lives in water, and then live their adult lives partially or completely on land.

2. **What are the three different orders of amphibians?**
   a. Gymnophiona, which is made up of worm-like amphibians called caecilians.
   b. Caudata, which is made up of newts and salamanders.
   c. Salientia, which is made up of frogs and toads.

3. **What are caecilians?**
Caecilians are worm-like in appearance. They do not have limbs and are usually without sight. They burrow into the decayed material on top of the soil and into the soil. They vary in length from 3 to 59 inches, and they have tentacle-like structures on their heads that enable them to find their way in their tunnels.

4.–5. **What are the characteristics of a newt?**

   What are the characteristics of a salamander?
Salamanders and newts make up the order Caudata, which includes those amphibians that have tails.
   Salamander is a broad term and can be used to apply to all members of the order Caudata. *Newt* is a more narrow term used when referring to members of the order Caudata that remain primarily aquatic throughout their lives.
   Salamanders and newts have lizard-like bodies with a tail. Most have thin, moist skin, but those living in dry areas have thicker, dryer skin.

6. **What are the differences between a frog and a toad?**
Frogs have slender bodies with long, powerful hind legs, short front legs, and webbed feet. They develop from tadpoles, and most species when fully grown can live in water or on land. Their skin is smooth. They are excellent jumpers and swimmers.
   Toads have short, thick bodies and short legs. The toad's skin is dry and has a rough appearance. True toads have skin glands that produce a poison, which can be very irritating to the mucous membranes of predatory animals. Most live on land in and around moist areas.

7. **Describe how to set up an aquatic, semiaquatic, and land habitat in an aquarium.**
Semiaquatic aquariums are set up for species that spend part of their lives in the water and part of their lives on the land. As tadpoles develop into small toads and frogs, they require semiaquatic habitats.
   To set up this type of habitat, an aquarium will need to be divided in half. A piece of glass or Plexiglas can be cemented into the bottom of an aquarium to divide it into two watertight halves. The divider should be about one-third the height of the aquarium. This again depends on the species being put into the aquarium.

The water half of the habitat can be set up the same as the completely aquatic habitat. To set up the land habitat, a 1-inch layer to small rocks or gravel should be placed in the bottom of the aquarium. A 1-inch layer of charcoal is placed over this; then, a 2- to 3-inch layer of sand or sandy loam is placed over the charcoal. On top of this, a layer of top soil is added so that it is level with the divider. Finally, leaves, branches, bark, and some plants are added to give the aquarium a natural look. The water in the semiaquatic habitat should be a dechlorinated. If that is not possible, the water should be allowed to sit in open containers for a couple of days so that the chlorine will evaporate. Water is added to the aquatic side of the aquarium until it is level with the land habitat. This allows the animals to move freely between the water habitat and the land habitat. The water should be changed every 2 or 3 days if a filtration and aeration system is not used. Plants and soil should be sprinkled daily with dechlorinated water so that the habitat will not dry out, and a desirable, damp environment is maintained.

The aquarium needs a cover to keep the animals from escaping; a piece of window screen or hardware cloth can be used. Temperature of the aquarium should be maintained at 60 to 80°F, depending on the species. A small light bulb can be added if heat is needed. Salamanders prefer temperatures around 60 to 70°F, whereas frogs and toads should be kept in an environment around 70 to 80°F. All amphibians require full-spectrum lighting. Full-spectrum bulbs are available from pet stores. These bulbs can emit full-spectrum light for only a limited time and should be replaced every 4 to 6 months, even if they have not burned out.

In preparing a land aquarium, proceed just as when preparing the land half of the semiaquatic aquarium. Even though it is a land habitat, a shallow bowl or dish of water should be provided. Amphibians will absorb the water through their skin. Make sure the bowl or dish is level with the soil so that the animals can crawl in and out freely. Remember that the plants and soil should be kept moist so that the damp environment is maintained. A small light bulb can be added if temperatures get too low.

CHAPTER 18

Reptiles

DISCUSSION QUESTIONS AND ANSWERS

1. **What is a reptile?**
Reptiles are cold-blooded vertebrates that possess lungs and breathe atmospheric air. They also have bony skeletons, scales, or horny plates covering their bodies and a heart that has two auricles and one ventricle.

2. **What are the four orders of reptiles?**
   a. Testudines (Chelonia), which is made up of turtles, tortoises, and terrapins
   b. Serpentes, which includes the snakes, pythons, and boas
   c. Squamata, which includes various species of iguanas and lizards
   d. Crocodilia, which includes the various species of crocodiles, alligators, caimans, and gharials

3. **What are the characteristics of turtles, tortoises, and terrapins?**
The term *terrapin* is applied to turtles found in freshwater and brackish water that are considered excellent eating.

   Turtles and tortoises have short, thick bodies encased in shells. All turtles and tortoises have limbs that enable them to walk on land. This group of reptiles has modified limbs that enable them to swim as well as to walk on land. Some turtles and tortoises spend most of their time in the water; their limbs are primarily adapted for swimming, and they move about on land with creeping and crawling motions. Turtles and tortoises do not molt their thick epidermal skin. A new epidermal scale forms beneath the old one as they grow. This allows for growth of the animal. These epidermal scales, or scutes, form rings that can be used to estimate the animal's age. However, this method may give an inaccurate indication, because more rings may form during a good year. The hard, tough layer of scales prevents moisture from dissipating from the body. In turtles and tortoises, this horny epidermal layer forms an exoskeleton, which has two parts. The upper part is called the carapace and also forms a part of the vertebrata and ribs of the animal. The lower portion is called the plastron.

   Turtles and tortoises do not have teeth. Their jaws form very sharp or crushing plates depending on the diet of the animal. The front part of the jaws form a horny beak.

4. **What are the characteristics of lizards?**
Lizards have long bodies and long, clearly defined tails. Their limbs are usually paired and attached to their bodies at right angles. This enables the animal to lift its body up off the ground when moving. The limbs of most lizards are weak, and some lizards have very short, stump-like limbs.

   The skin of lizards has a horny surface layer, which is formed by a hard, continuous covering of scales. These scales lie beneath the outer layer of the epidermis so the body can grow. As the animal grows, this outer layer is shed, allowing for further growth.

   Some lizard species do not have eyelids. Lizards have teeth that are fused to the jaw bones. Two species have teeth connected to poison glands.

   The tongues of lizards vary considerably. Some have short, fleshy tongues, which have very little movement, whereas others may have long, slender, and forked tongues.

The lungs of most lizards are the same size, but in some lizards and some snake-like lizards, the left lung is reduced in size, and the right lung is very long and lies between other internal organs. Lizards have three-chambered hearts.

5. **What are the characteristics of snakes?**
Snakes have long bodies that taper into tails. They do not have limbs, and their movement is a result of undulating movements of the body. The scales on the underside of the body project outward as the muscles are contracted and relaxed. These scales exert pressure on the surface and move the animal forward.

The skin of snakes has a horny surface layer that forms a hard, continuous covering of scales. These scales lie beneath the outer layer of the epidermis so that the body can grow. As the animal grows, this outer layer is shed, allowing for further growth.

Snakes cannot close their eyes. Instead of eyelids, they have a transparent layer, or brille, that permanently covers the eye. During the shedding process, this layer over the eye is shed and replaced with a new covering. Nocturnal species have vertical pupils that open very wide in dim light and close to small slits or pinholes in bright light. Diurnal species have round pupils.

Snakes have teeth that are fused to the jaw bones. Some snakes also have teeth fused to the palate bones. Some snakes have teeth connected to poison glands. All snakes have long, slender, forked tongues.

6. **What are the differences between pythons and boas?**
Boas are very similar to pythons except that they bear live young instead of laying eggs. Most boas live in Central and South America, although two species are found in North America.

7. **What are the characteristics of crocodiles, alligators, caimans, and gharials?**
Crocodiles, alligators, caimans, and gharials have long bodies and long, clearly defined tails. Their limbs are usually paired and attached to their bodies at right angles. This enables the animal to lift its body up off the ground when moving. The limbs of crocodiles and alligators are strong and powerful.

Crocodiles and alligators have bony, dermal scales covered by a horny epidermal layer, which must shed to allow for body growth. They have three-chambered hearts.

8. **Why do reptiles shed their skin?**
The skin of reptiles has a horny surface layer. In lizards and snakes, this horny layer forms a hard continuous covering of scales. As the animal grows, this outer layer is shed, allowing for further growth.

Crocodiles, alligators, and some lizards have bony dermal scales covered by a horny epidermal layer, which must be shed to allow for body growth.

9. **What are scutes? What are the different types of scutes?**
Turtles and tortoises do not molt their thick epidermal skin. Each year, a new epidermal scale is formed beneath the old one. This allows for the expanding growth of the animal. These epidermal scales are called scutes.

Scutes of the plastron are the gular, humeral, pectoral, abdominal, femoral, and anal scutes. Scutes of the carapace are the nuchal, supracaudal, marginals, centrals, and costals. Refer to Figure 18–2 in the text.

10. **Discuss the terms *oviparous, ovoviviparous,* and *viviparous*. What would be some advantages and disadvantages of each?**
Oviparous—animals that lay eggs.
Ovoviviparous—animals that retain eggs within their bodies until they hatch and then give birth to live young.
Viviparous—animals that retain the young within their bodies in a preplacental sac.

Oviparous reptiles lay their eggs in nest cavities in the soil. The eggs are covered and then left untended. When the young hatch, they are on their own, relying on inherited instincts for survival.

Snakes may be oviparous, ovoviviparous, or viviparous. Most are oviparous. Snakes that are ovoviviparous retain the eggs within their bodies until they hatch, and then they give birth to the young. The young receive no nourishment from the female, only the nourishment contained in the egg.

In a few species, the young are retained in a preplacental-type sac, which allows for some exchange of oxygen and nutrients. This is referred to as viviparous.

Some species of ovoviviparous and viviparous snakes protect their newly hatched young from predators.

Crocodilian females will stay within the vicinity of their nests to guard against predators. When she hears the sounds of the young coming from the eggs, the female will begin to dig the top of the nest away so the young can emerge.

11. **What are the four types of habitats that reptiles live in?**
    a. Terrestrial habitats for reptiles that live on land.
    b. Semiaquatic habitats for reptiles that live on land or in trees but spend part of their time in the water.
    c. Aquatic habitats for reptiles that live primarily in the water.
    d. Arboreal habitats for reptiles that live in trees.

12. **Discuss the types of housing and equipment that would provide the four habitats required by reptiles in captivity.**

    A land turtle will need a vivarium that is at least six times, in length and width, the length of the turtle's shell. If rocks, logs, or decorations are to be added, the vivarium should be larger. If additional turtles are going to be added to the vivarium, figure one-third more area needed for each additional turtle.

    For aquatic turtles, an area five times as long and three times as wide as the turtle's shell should be provided. Platforms made from wood or rocks should be provided so that the turtles can climb out of the water and bask in the rays of a heat lamp. The size of the basking area will influence the over-all size of the viviarium. The depth of the water in the vivarium should be at least 1 foot. The height of the sides should be sufficient to prevent the turtle from climbing out and escaping.

    The substratum in the bottom of a land turtle's vivarium can be 4 to 6 inches of soil, sand, dry leaves, or wood much. These materials allow for normal digging and hiding, but are difficult to clean. Lining the cage with newspaper, then providing a plastic hide box with organic bedding solves this problem. Misting the hide box and materials with water also provides an area of increased humidity. Depending on the species, branches, large rocks, and logs can be added. All land turtles should be provided with a shallow container of water. The depth of the water in the container should never be more than the height of the front edge of the turtle's carapace. The sides of the container should be gradually sloping so the turtle can easily get out of the water. No material or substratum is needed in the bottom of an aquatic turtle habitat.

    Small lizards, such as geckos and anoles, that are 5 to 6 inches long can be kept in vivariums a minimum of *that are about* 12 inches wide, 12 inches long, and 16 inches high. A vivarium of this size can keep one male and two females. The vivarium should also contain rocks, plants, tree limbs, or branches to give these active little lizards things to climb on. Some species are not social and should be housed individuals except during breeding.

    A ratio of length, to width, to height of 1:2:2 is sometimes used in determining the size of a vivarium for active tree- or wall-climbing lizards. In planning for terrestrial lizards, a ratio of 2:1:1 is sometimes used. Larger geckos and more active lizards will need a larger vivarium. A vivarium 60 inches long, 24 inches wide, and 24 inches high may be needed to keep several spiny lizards. Common iguanas will need a vivarium at least as large as 80 by 60 by 60 inches for one male and two females. Smaller vivariums would be needed if only one animal is to be housed.

    The most commonly used material for vivariums is glass aquariums. These should be fitted with an appropriate cover to keep the animals from escaping. Plexiglas can also be used to construct vivariums. Holes can be drilled in Plexiglas to provide ventilation. Glass and Plexiglas are definitely needed for aquatic, semi-aquatic, or rainforest habitats. Wood and wire mesh screen can be used to construct cages for dry habitats. Wood will deteriorate quickly if used for wet habitats.

    Water in aquatic and semiaquatic vivariums should be changed every 3 to 7 days to prevent the buildup of bacteria and harmful wastes. An alternative to changing the water is to install a circulation pump and filter system along with an aeration system to help maintain water quality.

Because reptiles are unable to generate their own body heat, a heat source must be provided. An incandescent light with a reflector shield will make a good light and heat source. The wattage of the bulb used will depend on the amount of heat needed and the distance the light is placed above the vivarium.

A thermometer and thermostat in the vivarium are important so that the temperature can be maintained in the preferred optimal temperature zone (POTZ) for the species. The temperature under the heat lamp may need to be as high as 100°F. To duplicate normal habitats, a timer may need to be added to a second heat lamp so that the lamp will be turned off at night, thereby maintaining higher temperatures during the day and cooler temperatures at night. Relative humidity should be maintained in the normal range for the species and can be monitored with a hygrometer.

Sunlight is important not only as a heat source but because the skin absorbs the ultraviolet (UR) rays that are needed in calcium metabolism, formation of pigment, and vitamin D synthesis. Inadequate lighting can lead to metabolic bone disease and other ailments. A full-spectrum UVB fluorescent lamp is recommended. A UVB heat lamp will provide the needed UV rays as well as heat. The lamps should not be placed on glass, because the glass will filter out the UV rays. These lamps emit full-spectrum light for a limited time and should be replaced every 4 months. In warm months, house the reptile outside during the day. Outside caging should be predator proof and out of full sun.

Vivariums for aquatic and semiaquatic species need a method of heating the water, which should be maintained at a temperature mimicking the species' natural environment. Aquarium heaters can be used. To prevent reptiles from moving or damaging the heating unit, it should be placed between a couple of clay bricks.

If additional heating is needed, a heating cable or a heating mat can be used. Heating cables can be run through the substratum in the bottom of the vivarium. A piece of wire mesh screen placed over the cables will prevent reptiles from digging up the cables and possibly causing burns. Placing heat cables or mats under the bottom of the cage and varying the thickness of the substrate can increase safety. Heat rocks are not recommended because many animals have been burned by them.

Reptiles that originate in tropical rainforests may need to have sprinkler or misting systems to duplicate the daily rainfall of their original habitats; these will need to be on a timer. Heat and humidity requirements for this type of habitat will differ considerably from other types of habitats.

13. **What is an ectotherm?**
An ectotherm is an animal that cannot generate its own body temperature and will take on the temperature of its environment.

14. **Discuss the types of food needed to maintain reptiles in captivity.**
Turtles should be fed every day. All aquatic turtles are primarily carnivorous or omnivorous. In captivity, aquatic turtles eat commercial turtle pellets, lettuce, raw fish, and worms.

Land turtles are primarily herbivorous, feeding on grasses and plants. They should be given pieces of alfalfa, clover, or timothy hay; leafy, dark greens like kale; carrots; sweet potatoes; green beans; radishes; strawberries; and cantaloupe. Omnivorous turtles can be fed these and offered worms; soaked, dry, low-fat dog food; or cooked chicken in small amounts.

All snakes are carnivorous. Small snakes in captivity will readily feed on small fish, earthworms, mealworms, crickets, and other insects. Larger snakes will feed on baby mice and rats. They may also feed on frogs and toads, if available. The big snakes will feed on full grown mice, rats, baby chicks, guinea pigs, chickens, and rabbits.

If several types of food have been offered but not eaten for 4 weeks or more, seek advice from a consultant.

Feeding lizards in captivity is fairly easy. Most are insect eaters, and crickets, mealworms, and earthworms are readily available. Iguanas and larger lizards will eat mealworms, earthworms, lettuce, flower blossoms, fresh fruit, vegetables, ground meat, and dog or cat food.

15. **What are the signs of good health to look for when obtaining a reptile?**
The reptile should be observed carefully, making sure it has been eating or is eating. An animal that has been eating and is well nourished will be filled out, and its ribs, vertebrae, and pelvic bone will be covered and not

show. The animal's eyes should be open, and the animal should be alert to its surroundings. Its mouth should be closed.

16. **What are some signs of ill health?**
Any frothing or foaming at the mouth or around the nose, or running at the eyes, would indicate an unhealthy animal. If the animal shows bite wounds or marks of other injuries, these should be healed and scarred over. One should avoid purchasing animals with open wounds, wounds that are seeping, or wounds that do not appear to be healing. A turtle that is sick will usually drag the rear of its shell on the ground instead of carrying the shell horizontally off the ground. Aquatic turtles that are sick may lean to one side when in the water. Look carefully for parasites. Avoid animals that appear to be heavily infested.

CHAPTER 19

Birds

DISCUSSION QUESTIONS AND ANSWERS

1. **What is a bird?**
Birds are two-legged, egg-laying, warm-blooded vertebrates with feathers and wings, and they are believed to have evolved from prehistoric reptiles.

2. **Why do many people believe that birds evolved from prehistoric reptiles?**
Evidence includes that birds have scales on their legs and feet similar to those on reptiles, birds are egg-laying like many species of reptiles, and birds have feathers, which are modified scales. Both birds and reptiles have similar soft anatomy, including musculature system, brain, heart, and other organs. Birds and reptiles show many skeletal similarities. Bird-like fossils have been discovered in several countries.

3. **What are the types of feathers found on birds, and what are their purposes?**
Contour feathers give the bird its outward form; those contour feathers that extend beyond the body are used in flight and are called flight feathers. The small, soft feathers located under the contour feathers are referred to as down feathers; their primary function is to conserve body heat. Short, hair-like feathers called filoplume feathers are found over the birds' entire body. They play a key role in sensory function and serve as insulation. They are believed to send information about how feathers need to be adjusted to keep them in place during preening, display, and flight. These filoplume feathers appear as the hairs on chickens that have been plucked of their contour and down feathers. Powder-down feathers have tips that break down as they mature and release a talc-like powder that provides waterproofing and luster; these feathers are found on parrots and cockatoos.

4. **What are the three orders of birds discussed, and what are the main characteristics of each?**
Psittaciformes—cockatoos, macaws, conures, parrots, lories, and parakeets
Piciformes—woodpeckers and related species (toucans)
Passeriformes—perching birds

    Psittaciformes are the parrots and related species; the order consists of only one family, Psittacidae, and six subfamilies. All members of this order have similar hooked bills and peculiar mallet-shaped tongues. The hooked bills are powerful and are used to obtain food and as weapons against enemies. These birds have strong, heavy legs and feet that are used in climbing through trees. The hooked bill can also be used in climbing. Most species have harsh voices and the ability to mimic.

    Piciformes consist of woodpeckers and related birds. There are six families in the order. Most species in this order are strictly arboreal, rarely descending to the ground. They feed on fruits, vegetation, and insects. Only the toucans are kept as pets.

    Passeriformes include the perching birds. There are 65 families in the order. Perching birds have three toes that point forward and one toe that points backward; the toes are not webbed. Perching birds are also noted for their pleasing songs.

5. **Name three nutritional ailments of birds, and list the symptoms of each.**
   - Iodine deficiency—swelling of thyroid glands
   - Calcium and phosphorous ailments—softening of bone, lameness, stiff-legged gait, constant resting position
   - Candidiasis—regurgitation of food, lack of appetite, general signs of illness

6. **What are the guidelines to follow when placing a cage in a room?**
   The cage should be placed in an area of the room that is free from drafts, where constant temperature is maintained, not in direct sunlight, and away from poisonous plants or other hazards the bird can reach.

7. **What are some important factors to remember when selecting a cage?**
   a. The cage should be big enough for the bird or birds.
   b. The cage should be easy to clean.
   c. The cage should be large enough to provide room for feeders, waterers, nest boxes, and other items.
   d. The cage should be affordable.
   e. The cage should be made of stainless steel or anodized aluminum or be chrome-plated.
   f. Galvanized wire contains zinc, and should be avoided.

8. **Describe how a cage for large, parrot-type birds should be constructed.**
   Equipment used for the large parrot-type birds must be constructed of very heavy materials; wood and plastics cannot be used for parrot cages because they will shortly be destroyed. Heavily constructed metal cages must be used. The cage should be inspected for sharp points or edges that can cut or injure the bird's tongue or feet. Secure latches or locks should be used.

   If one chooses to construct a cage for macaws, the wire size should be ½ × 3-inch wire mesh screen in 12- to 14-gauge wire. Smaller mesh wire may be used for outside cages if mice, rats, or other birds are a problem.

   Wood makes the best perches. Parrots will soon destroy wood perches, and the perch will have to be replaced. Wood does give the parrot the opportunity to exercise its beak, which helps keep the beak trim and the bird busy.

   Feed containers for the large parrot-type birds must be constructed of sturdy materials. Glass, ceramic, or stainless steel are preferred.

9. **What are two types of seeds consumed by birds, and what are some examples of each?**
   Cereal seeds are canary seed, millet, corn, and dehusked oat kernels. Oil seeds are sunflower seeds, peanuts, safflower, pine nuts, rape seeds, maw, niger, and linseed.

10. **Why is grit important in the diet of some birds? Which birds do not need grit?**
    Birds cannot grind up their food. They can break the seeds open with their beaks, but the actual grinding of food takes place in the ventriculus. For some species of birds this is accomplished with the aid of grit that must be supplied in their diets. Grit is available from pet stores in two forms: soluble and insoluble.

    Mynah birds will not eat seeds and do not need grit or cuttlebone in their diets. Lories and lorikeets feed on nectar and pollen. They will also eat fruit. Lories and lorakeets do not need grit in their diets.

11. **Describe the steps in getting a bird to feel comfortable with its handler and the steps in training.**
    After bringing a new bird home, begin trying to offer the bird a treat through the bars of the cage. The bird may refuse the treat, but one should be patient and continue at regular intervals to offer a treat. One should talk softly to the bird, and soon the bird will take the treats. The next step is to open the door of the cage and offer the treat through the open door. One should move slowly and talk calmly to the bird. The bird may retreat, but again, one must be patient and the bird will soon be taking the treats. Next, a stick or perch should be introduced into the cage and the bird encouraged to step up onto this new perch. The perch should be pressed up against the bird's chest just above its legs. Parrots use their beaks to aid in climbing, so one should not be alarmed if the bird grasps the perch with its beak. The feet should step up on the perch next. After the bird becomes accustomed to climbing on a perch held into the cage, one can slide a hand or finger

(depending on the size of the bird) in under the stick and attempt to get the bird to perch on the hand or finger. While attempting this, one should offer the bird a treat to distract it and continue to talk calmly to the bird. A leather glove may be needed when working with the large parrot-type birds. Once the bird feels comfortable perched on the hand, one can try to remove the bird from its cage while perched on the hand. The bird may jump from the hand at first, but one should keep trying; this process may take several days. Some birds are more adventurous than others. Patience is important. The next step is to try to get the bird to use a perch outside its cage or to perch on one's arm or shoulder.

12. **What are some signs that might indicate poor health in a bird?**
    a. Sick birds will sit on the perch with one or both wings drooped and their tails pointed downward.
    b. Ill birds stand unsteady on their perches and they are restless, shifting weight between legs or favoring one leg.
    c. Runny feces could indicate a digestive ailment.
    d. Fluffed feathers could indicate that the bird is trying to retain body heat; the bird may be very ill.
    e. Lack of activity or failure to fly may indicate an ailment.
    f. Eyes that are continually closed or have discharges are signs of ailments.
    g. Irregular or difficult breathing, wheezing, or noisy breathing could indicate a respiratory ailment.
    h. Loss of appetite or failure to eat may indicate an ailment.

13. **What are some external parasites of birds, and what are the symptoms that might indicate a bird has external parasites?**
    - Red mites, *Dermanyssus gallinae*—restlessness, scratching and picking
    - Feather mites, *Knemidocoptes laevis*—restlessness, severe scratching, feather picking, skin irritation
    - Scaly face and leg mites, *Knemidocoptes pilae*—white, scaly deposits

14. **What are some internal parasites of birds, and what are the symptoms that might indicate a bird has internal parasites?**
    - Roundworms—intestinal blockage, poor plumage, weight loss
    - Tapeworms—small, rice-like, white segments in feces
    - Flagellated protozoan, *Trichomonas gallinae*—small, yellowish lesions on mouth and throat linings

15. **What are some bacterial diseases of birds, the symptoms, and the treatments for them?**
    - Psittacosis (parrot fever or chlamydiosis)—nasal discharges, listlessness, labored breathing
    - Colibacillosis—foul-smelling discharge
    - Pasteurellosis, fowl cholera—fever, depression, anorexia, ruffed feathers
    - Bumblefoot—swollen joints, difficulty walking
    - Tuberculosis—slow weight loss

16. **What are some viral diseases of birds, the symptoms, and the treatments for them?**
    - Psittacine beak and feather disease—lack of new feathers, soft beak and nails
    - Newcastle disease—depression, loss of appetite and weight, sneezing, head bobbing
    - Pacheco's parrot disease—sudden death or brief period of anorexia, depression, diarrhea

CHAPTER 20

Fish

DISCUSSION QUESTIONS AND ANSWERS

1. **Describe the characteristics of fish.**
   Fish are cold-blooded vertebrates that breathe with gills and move with the aid of fins. They are the most numerous vertebrates, with more than 30,000 species. Most fish are covered with scales, which are thin, bony plates that overlap each other and provide protection.
   The skin contains chromatophores, pigment cells that give the fish its colors. The colors a fish has usually allow it to blend with its surroundings, and most fish are able to change their colors if necessary. Sensory receptors are also contained in the skin.
   Fins are movable structures that aid fish in swimming and maintaining balance. Most bony fish have rayed fins.
   Fish breathe through organs called gills. Water is drawn in through the mouth by the constant opening and closing, forced back into the pharynx, and exited out through the gills.
   Most bony fishes have swim bladders in their abdominal cavities. The swim bladder is filled with gases produced by the blood that enable the fish to maintain itself at a particular depth.
   The fish heart consists of two chambers: the atrium and ventricle.
   Most fish have lateral lines composed of a series of pressure-sensitive cells. These cells, or neuromasts, are contained in tubes along the lateral line.
   The digestive systems of fish vary depending on the type of food consumed.
   The eyes of fish are similar to those of other vertebrates but differ in a couple of ways. The fish's eye has a spherical lens that focuses by moving within the eyeball, not by changing the curvature of the lens. Fish do not have eyelids; the eye is kept moist by the flow of water.
   All fish have inner ears. Species that have swim bladders have a more acute sense of hearing because the bladder acts as a resonator and amplifies the sound.
   Fish have taste buds in their mouths, on their lips, and on their bodies and fins. Their senses of smell and taste are highly developed. Some fish have taste buds on their barbels, which are whisker-like projections around their mouths.

2. **What are three main classes of fish, and what are their important characteristics?**
   - Class Agnatha or Cyclostomata includes the jawless fish, possessing sucking or filter-feeding mouths.
   - Class Chondrichthyes includes fish with cartilaginous skeletons.
   - Class Osteichthyes includes fish with bony skeletons.

3. **Describe the general characteristics of the 12 orders that are commonly found in the aquarium trade.**
   The Order Cypriniformes is made up of three suborders, 19 families, and 3,500 species. Eight families are common in the aquarium trade. This order includes fishes that are covered with cyloid scales.
   The Order Siluriformes is made up of 24 families and more than 2,500 species; they are commonly referred to as catfish. Most species are covered with skin but lack scales. Some species are covered with bony

plates; most species have adipose fins. The pectoral and dorsal fins have stiff, sharp rays or spines. Their bodies are usually flattened on the underside, and they have barbels; these are characteristics that aid in their bottom-dwelling, nocturnal habits.

The order Atheriniformes is made up of three suborders and 15 families. Fishes in this order include those with extended lower jaws, the annual fish, small fish with modified anal fins, fish with specialized eye structure, and those that lack a lateral line.

The order Perciformes consists of more than 6,000 species in 16 suborders and approximately 160 families. Species in this order do not have their swim bladders connected to open ducts to their throats. The dorsal, anal, and pelvic fins usually have spines. The dorsal fin consists of an anterior (front) part that has spines and a posterior (rear) part that has soft rays.

The order Tetraodontiformes is represented by the family Balistidae. This family is made up of a group of saltwater species referred to as triggerfish. They each have three dorsal spines, with the second locking the first into an upright position.

The order Mormyriformes is represented by the family Mormyridae, which contains 130 species. They are referred to as mormyrs and elephant-snout fish. The lower lip is elongated into a tool that is used to dig for food on the bottom of streams. They also emit electrical fields around their bodies that enable them to move around in the darkness.

The order Osteoglossiformes is represented by the family Notopteridae and contains six species consisting of fishes with long anal fins and reduced or no dorsal fins. They move about by the undulating motions of their long anal fins.

The order Scorpaeniformes is represented by the family Scorpaenidae and is made up of the lion fish, also called dragon fish or turkey fish. The fin rays are long, and the spiny ray is poisonous.

The order Gasterosteoidei is represented by the family Syngnathidae and consists of the seahorses. They have hard body coverings and no caudal or anal fins; they swim in a vertical position. Seahorses have prehensile tails that they can use to anchor themselves to plants and coral.

4. **What are the three types of filter systems, and how does each one work?**

The purpose of the filter is to remove solid waste and uneaten food materials from the water. Filter systems work in one of three ways: through the mechanical removal of waste and food materials, by chemical removal of dissolved materials, and by biological filtration or convection of harmful substances into harmless ones.

Older aquariums used internal mechanical filters driven by airstones. As the air rose, this pulled the water through the media in the box. The media separated the waste and food materials out of the water, and the water was returned to the tank. Newer models still use the air-driven uplift principle, but instead used external box filters or internal sponge filters.

Most aquariums today use a power filter with an electric motor connected to an impeller by a magnetic device through the wall of the filter box. By not housing a drive shaft through the filter box, there is less chance for leaks in the system. A power filter increases water flow through the filter media and removes more waste materials. It is important that the connection of the return hose to the aquarium be secure. If the hose is loose, large amounts of water can be pumped onto the floor, and the entire tank could possibly be emptied. Undergravel filters and internal and external canister filters are now available with the use of electric power pumps.

There are at least a half-dozen different types of chemical filtration units on the market today. One method of chemical filtration is accomplished with the use of activated charcoal; the charcoal soaks up dissolved minerals and chemicals. Activated charcoal can be placed in box filters, or return water can flow through the charcoal. The charcoal must be replaced periodically as it reaches a point where it can no longer absorb dissolved material.

Aeration is achieved by the use of electrically driven air pumps; pumps are either a vibrator-diaphragm type or a rotary-vane type. Pumps are available in several sizes; the size needed depends on the size of the aquarium, the number of airstone filters, and other equipment used.

Biological filters neutralize toxic substances, especially ammonia excreted from fish during respiration and produced from decaying waste and food materials. In this system, a slotted plastic plate is placed in the bottom of the aquarium. A 2- or 3-inch layer of gravel is placed over the slotted plate; the gravel should be 0.125 inch in diameter. Gravel that contains large particles allows food to fall into the spaces around the gravel and decay; too-small particles block the action of the air pump. Marine aquariums should have coral sand placed above the slotted plate. In some aquariums, peat should be placed over the slotted plate. The slotted plate has an uplift tube on the corner; this tube should be placed at the back of the aquarium. When the tank is filled, an air line with an aerator attached is placed down the uplift tube. This line is connected to an air pump. Air is pumped down the line and through the airstone. As air bubbles rise in the uplift tube, water is drawn through the gravel, up through the uplift tube, and back into the aquarium at the surface. The gravel layer acts as a filter for suspended particles in the water. After several hours of aeration, colonies of aerobic bacteria (bacteria that use oxygen) will begin to grow in the gravel; this process is called nitrification.

5. **Describe the nitrogen cycle that takes place in an aquarium.**
Nitrosomonas bacteria convert ammonia into nitrites, and then *Nitrobacter* bacteria convert the nitrites into nitrates. Nitrates may inhibit the growth of fish, but they are not nearly as toxic as ammonia. Nitrites and nitrates may be utilized by plants in the aquarium.

6. **What is a power filter, and what is the advantage of using one?**
A power filter uses an electric motor connected to an impeller by a magnetic device through the wall of the filter box. By not housing a drive shaft through the filter box, there is less chance for leaks in the system. Power filters increase water flow through the filter media and remove more waste materials.

7. **What are some common foods that fish consume?**
Flake foods are ideal for small fish up to 4 or 5 inches long, if appropriate. These foods are produced from the meat of fish, fish eggs, wheat, and vegetables. They usually contain additional vitamins and minerals. These food materials are also processed into pellet form for larger fishes; floating fish sticks are for large, top-feeding fish, while small pellets that sink slowly are for middle-feeders, and sinking tablets are for bottom-feeders.

Several live foods are also available; among these are small crustaceans called water fleas, *Daphnia*. Brine shrimp, *Artemia*, are small shrimp that live in salt lakes or brackish waters. These shrimp should be rinse off and may be fed to both small and large fish. River shrimp and bloodworms are available for still larger fish. Earthworms, flies, maggots, wood lice, caterpillars, crickets, and grasshoppers are other live foods that can be used. One problem with feeding live foods is the parasites and diseases that they may carry can be harmful to fish.

Freeze-dried and frozen foods are basically the same as live foods, except they are in a safe form. Some freeze-dried or frozen foods available are *Mysis* shrimp, Pacific shrimp, tubiflex worms, krill, and plankton.

Carnivorous fish can be fed minced or chopped meat; common foods include beef heart, liver, raw fish meat, and shellfish meat. Pieces of turkey and chicken can also be used. Meat pieces can also be thinly sliced and frozen; these can then be broken and fed. Some of the larger carnivorous fish may need to be fed other fish; small goldfish may be used for this purpose.

Many fish also require vegetative material in their diets. Chopped or shredded lettuce, chopped spinach leaves, canned peas, wheat germ, and oat flakes can be used. Any vegetables not consumed within 8 hours should be removed.

Marine species of invertebrates are of four types:
a. Those that feed on plankton that is filtered from the water, including stone and horny coral, tube worms, bivalves, some species of sea cucumbers, and crustaceans. Commercially prepared plankton foods and frozen foods are available for the plankton feeders.
b. Those that feed on plant material, including sea urchins, mollusks, and sea slugs. A diet of lettuce and spinach will hopefully prevent them from feeding on aquarium plants.
c. Those that are carnivorous, including crabs, starfish, sea anemones, shrimp, and lobsters. They can be fed small pieces of crab, fish, shrimp, and flake foods. Sea anemones should be fed only when their tentacles are out (in bloom). Drop the food onto their tentacles.

d. Those that are scavengers, including sea cucumbers. They feed on debris and uneaten foods on the bottom of the aquarium.

8. **What are the five types of egg-laying fish, and what are their habits?**
   a. Egg-scatterers are fishes that lay their eggs in a haphazard manner on the floor of an aquarium. Some species lay adhesive eggs that stick to the gravel on the aquarium floor or on decorations and plant materials. Other species lay nonadhesive eggs. There is no parental care of the eggs or young; when the eggs are laid, they are forgotten. In many cases, steps must be taken to prevent the parents from eating the eggs and the newly hatched fry.
   b. Egg-buriers are fish that lay their eggs in the mud of rivers and ponds. The adults lay their eggs in the mud and die when the rivers and ponds dry up. The eggs survive in the mud, and when the next rains come, the young hatch. In aquariums, they lay their eggs in the material on the floor of the aquarium.
   c. Egg-depositors have complex spawning routines and are excellent parents; these fish will usually select their own mates. They clean off a nesting site where the female deposits her eggs and the male fertilizes them. The parents usually take turns guarding the nest; they constantly fan the nest and eggs with their fins to keep them clean from dirt or silt and dust that may settle on them. When the young fry hatch, the parents will keep watch for several days to protect them from predators.
   d. Mouth-brooders are fish that carry their eggs in their mouths until they hatch, after which the young may continue to be carried in the parent's mouth until they are ready to fend for themselves. In some species, the young, after leaving the parent's mouth, may return if threatened. After the eggs deposited by the female are fertilized by the male, the female, and in some species the male, will go around the pick up the eggs by mouth. During this incubation period of 2 to 3 weeks, the parent does not eat.
   e. The nest-builders construct nests in which the eggs are deposited; it is usually constructed by the male. This nest may be a bubble nest on the surface made from saliva-blown bubbles or may be prepared from materials found on the floor of the aquarium. After the eggs are fertilized, the female should be removed because the male will become aggressive toward her; the male then guards the nest.

9. **What are signs that might indicate a fish has a disease or other ailment?**
   a. Unusual behavior such as swimming movements
   b. A fish leaning to the side, floating to the surface, or sinking to the bottom
   c. Fins not extended and laying flat against the body
   d. Caudal fins closed or rolled
   e. A fish shying away from the shoal
   f. A fish slower in its movements than others
   g. Respiration faster and deeper than normal
   h. Fish up on the surface, gasping for air
   i. Fish that are rubbing against decorations on the bottom
   j. Fish that are not eating
   k. Fish that are thin with sunken sides
   l. Abnormally large or swollen belly
   m. Abnormal color
   n. Frayed fins
   o. Malformation of the back and spinal column
   p. Cloudiness of the eyes
   q. Scales sticking out away from the body instead of lying flat

10. **What are common diseases caused by parasites, and what are the symptoms and treatments for each?**
    - White spot or "Ich"—white spots on body and fins
    - Slime disease—large amounts of mucus
    - Hole-in-the-head disease—large, open sores around head, dorsal line, and dorsal fin

- Velvet disease and coral disease—appears as velvet-like, gold coating on sides
- Fungus growths—white fungus growths on fins, mouth, eyes, and gills

11. **What are two species of flukes that attack aquarium fish, and what are the symptoms and treatments for each?**

    Two species of flukes that can cause problems are *Dactylogyrus* and *Gyrodactylus*. Resulting damage from both can be fatal depending on the degree of infestation. Flukes are tiny, worm-like parasites that attach themselves to the gills and body of a fish. They have numerous hooks on the rear part of their bodies and sucking mouths. A single fluke has both male and female reproductive organs (hermaphroditism). *Dactylogyrus* permanently attach themselves to the gills. Fish affected with *Dactylogyrus* will have rapidly moving gills, swim at the surface, and pant heavily, trying to obtain more oxygen. Their gills will be covered with slime, and parts of the gills may be eaten away. Fish may also scrape their bodies up against objects trying to relieve the irritation. *Gyrodactylus* primarily attach themselves to the body of the fish. The infected fish will also scrape up against objects, their color will fade, the fish will produce more slime in response to the irritation, and the fins may also become ragged and eaten away. Commercial treatments are available at pet stores with fish supplies. Infected fish should be removed and treated in a hospital aquarium quarantine. The aquarium that the fish were removed from should be thoroughly cleaned before returning fish to it.

12. **What causes finrot, and how can it be controlled?**

    Finrot is a condition especially prevalent in brightly colored fish or those with long, trailing fins, such as the black mollies. The edges of the fins will start to lose their color, the tissue between the rays will begin to break down, and the fins will become ragged. The fins become shorter as the condition persists. The disease is believed to be caused by a bacterium, along with a contributing factor such as a vitamin-deficient diet, poor water conditions, or fin nipping by other fish. Commercially prepared treatments are available; treatment of marine species may be more difficult.

CHAPTER 21

Hedgehogs

DISCUSSION QUESTIONS AND ANSWERS

1. **What is the genus name of the common hedgehog?**
   The common hedgehog or white-bellied hedgehog is of the genus *Atelerix*.

2. **What things should be considered when buying a hedgehog for a pet?**
   The amount of time one has and one's ability to care for a hedgehog.
   One can be pricked by the spines of the animal when handling it.
   Check the cage of the animal to make sure the cage and animal are clean.
   Check the health of the animal.

3. **What are the major identifying characteristics of the hedgehog? What are some other characteristics of the hedgehog?**
   The most notable characteristic of hedgehogs are the spines that the animal has developed as a means of defense.

   Other characteristics are: long snouts, four digits on each foot, and 36 to 48 teeth. They have large eyes and ears. Their tails are very short and most species have short legs. Their life span is about 5 years. They make very little noise except for low grunts and clicks.

4. **What is meant by the term *anointing*? What are the possible reasons for this behavior?**
   When the animal comes across a new smell or object, it licks at the smell or object, producing a flow of foamy saliva. The animal will then stiffen its front legs and swing its head from side to side. Using its tongue, it spits and smears its spines with the saliva.

   The reason for this behavior is not known. It may be a form of protection from predators, or a method of attracting a mate.

5. **What are the common foods for hedgehogs?**
   Hedgehogs are insectivores and nocturnal. They feed on insects, worms, snails, bird eggs, and small invertebrates such as frogs and snakes.

   Foods formulated for hedgehogs are now available in some localities. These would be the preferred diet. If these formulated foods are not available, good-quality cooked beef and poultry cut into small pieces are favored. Other foods that one can try feeding are dry kitten or cat formulas. The commercial foods can be supplemented with some small pieces of fruits and vegetables. Hedgehogs love mealworms, wax worms, and crickets. Feed the worms and crickets primarily as treats, two or three at a time, every 2 or 3 days.

6. **How should a hedgehog be handled?**
   A hedgehog not familiar with you or your scent will roll up into a ball when you get near it. This is the animal's natural reaction to danger. The spines on the hedgehog are very sharp. When one attempts to pick up an animal that has rolled up into a ball, it is possible to get some painful pricks from the spines. Gloves would be recommended for handling hedgehogs that are rolled up into a ball. As an animal gets familiar with

59

its handler and his or her scent, it will not sense any danger and will not roll up into a ball. The hedgehogs can then be picked up with one hand on each side of its body and under the belly. A hedgehogs that is familiar with its handler will many times walk up into one's cupped hands.

7. **What are some common diseases and ailments of hedgehogs?**
A change in eating habits, such as reduced appetite, drinking excessive amounts of water, or a lack of thirst, may be the first sign of illness in a hedgehog. Dark, torpedo-shaped fecal material is the sign of a healthy animal. Soft, green, or liquid fecal material may indicate a problem.

Poor hygienic conditions cause many of the hedgehog's health problems. Following are some ailments that affect hedgehogs:
   a. Mites
   b. Intestinal threadworms (*Capillaria* species), intestinal worms (*Brachylaemus* species), and coccidia (*Isospora* species)
   c. Eye infections
   d. Diarrhea
   e. Overgrown toenails

8. **Define the terms *hibernation* and *estivation*. How are they similar and how do they differ?**
Hedgehogs in the wild that are found in cold climates survive the winter's cold temperatures by going into a state of hibernation. During hibernation the animal's heart beat drops by 90%, its body temperature declines and respiration is reduced. This is a tremendous savings in energy requirements. The animal is sustained during hibernation by fat reserves that it stores in its body during warmer weather.

Estivation is similar to hibernation but not as intense. Hedgehogs that are found in hot, dry climates and reduced food supplies will go into a state of estivation. The animals seek out a cool hiding place. An animal in a state of estivation will have a reduced metabolic rate but not to the low levels that occur with animals in hibernation. They may wake and exit their hiding place on a regular basis.

CHAPTER 22

Sugar Gliders

DISCUSSION QUESTIONS AND ANSWERS

1. **What are the countries of origin for sugar gliders?**
   Sugar gliders are native to northeastern regions of Australia, Tasmania, Indonesia, and Papua New Guinea.

2. **What is a marsupial?**
   Marsupials have a pouch called a marsupium in which the young are raised during early infancy.

3. **What are the differences between marsupial mammals and placental mammals?**
   After young marsupials are born, they will travel up their mother's belly and into the pouch. Once inside the pouch, the young will attach themselves to a nipple, where they are nourished. The young remain in the pouch until they are strong enough to venture outside the pouch. They will return to the pouch for warmth and nourishment.
   Placental mammals do not have a pouch in which the young are raised.

4. **Describe the stages in the birth and growth of young sugar gliders.**
   The gestation period is about 16 days. The young gliders, called joeys, are only about two-tenths of an inch long. After being born, the young gliders will crawl their way into the mother's pouch, where they take hold of one of four nipples. The nipple swells in the young glider's mouth so that it is securely attached. It will be about 2 weeks before the young are large enough to be detected in the pouch. There is usually one or two in the litter. The young gliders will remain attached to the nipples for approximately 30 to 40 days and remain in the pouch for another 30 days. At about 60 to 70 days (86 days after conception), the young gliders will leave the pouch, returning occasionally for warmth, security, and to nurse. At this point, they are fully furred and their eyes are open. Weaning takes place at about 105 to 125 days (121 days after conception). Young gliders will become sexually mature at about 8 months of age.

5. **Describe the identifying characteristics of sugar gliders.**
   One of the most distinguishing characteristics of the sugar glider is the patagium, or thin, furred membrane that stretches from the wrists to the ankles. When not gliding, the membrane appears as excess skin on the side of the animal.
   The female sugar glider has a pouch. This pouch can be seen as a small slit at the lower abdominal area.
   Sugar gliders have a body length of 5 to 6 inches with a tail of near equal length, giving the animal a total length of approximately 10 to 12 inches. They weigh 3 to 5 ounces, with the males being slightly larger than the females.
   The head is somewhat triangular. The muzzle is short and the nose is rounded. Their eyes are large and round. Sugar gliders have 40 to 46 teeth, including the two incisors that are large and protrude forward.
   Sugar gliders normally have soft, gray-colored fur with cream-colored undersides. They have a black stripe that runs from between the eyes, over the head, down the center of the back, to the base of the tail.

Sugar gliders have black rings around the eyes and black stripes on the front of each leg. They have a bushy tail, which is also gray in color except for the last couple of inches, which is black. Their ears are furless and fairly large. The ears move independently of each other and are constantly in motion, trying to pick up sounds.

Adult males have a scent gland that appears as a bald spot on the head, and this bald spot separates the black stripe as it crosses over the head.

6. **What should one look for when selecting a sugar glider for a pet?**
    First, one should check with state and local laws to make sure they can be kept legally.
    One should plan on getting more than a single glider.
    When one is selecting individual animals, choose recently weaned gliders.
    Other things to look for when selecting individual gliders are:
    a. The eyes should be bright, clean, and clear.
    b. The animal should show signs that it is eating well.
    c. There should be no running or discharges from the nose.
    d. The fur should be soft, thick, and clean.
    e. There should be no signs of diarrhea.
    f. The animal should be active and alert.
    g. The animal should be friendly and curious.

7. **Describe the proper environment for a sugar glider.**
    Sugar gliders, being arboreal in the wild, are very active and very curious. They like to climb and explore. When selecting the cage for these animals, it needs to be large enough to allow them to explore, climb, exercise, and glide. The larger the cage, the better suited it is for gliders. Tree limbs or branches should be added to the cage to give the cage a more natural environment.

    Sugar gliders should be supplied with toys to keep them occupied. Toys used for birds such as swings, perches, ladders, and bells that are made well will be enjoyed by gliders. Gliders can also be supplied with pieces of PVC pipe or cardboard paper towel rolls. Nest boxes made of wood or plastic can also be used. These should be placed high in the cage where the gliders will feel more safe and secure.

8. **List some of the suggested foods that can be fed to sugar gliders.**
    Commercial diets for sugar gliders are available and should be included as part of their diet. The diet should consist of approximately 75 percent fruits and vegetables and 25 percent protein such as meats and insects.

    A list of suggested fresh fruits includes apples, apricots, bananas, melons, cherries, dates, figs, grapes, oranges, peaches, pears, pineapple, plums, and strawberries.

    A list of vegetables that one may feed includes asparagus, beans, beets, broccoli, carrots, cauliflower, peas, sweet potato, and sweet corn.

    Lean, cooked beef and poultry are good sources of protein. Hard boiled eggs, chopped or mashed, and scrambled eggs are good protein sources and may be fed occasionally.

    The sugar glider will enjoy crickets, grasshoppers, mealworms, and wax worms. These can be fed as treats. Feed live food sparingly because some insects are high in fat and low in calcium.

9. **What is stress, and what are conditions that can lead to stress?**
    (See Glossary.) Stress is any condition that is not normal, and can be caused by putting an animal in a situation or under conditions that it is not used to.

    Because sugar gliders are nocturnal, they should be kept in a fairly quiet area and allowed to sleep during the day. They can be easily stressed if awakened and taken out of their cages in daytime hours.

ANSWER KEY FOR STUDENT WORKBOOK

SECTION 1

CHAPTER 1
Introduction to Small Animal Care

1. (a) cretaceous
   (b) Carolus Linnaeus
   (c) binomial nomenclature
   (d) taxonomy

2. kingdom, phylum, class, order, family, genus, species

3. Characteristics of mammals:
   (a) give birth to live young
   (b) produce milk
   (c) have a placenta to nourish the growing embryo
   (d) have hair

4. (a) duck-billed platypus
   (b) echidna

5. Crossword Puzzle "Small Animal Care"

Crossword Small Animal Care

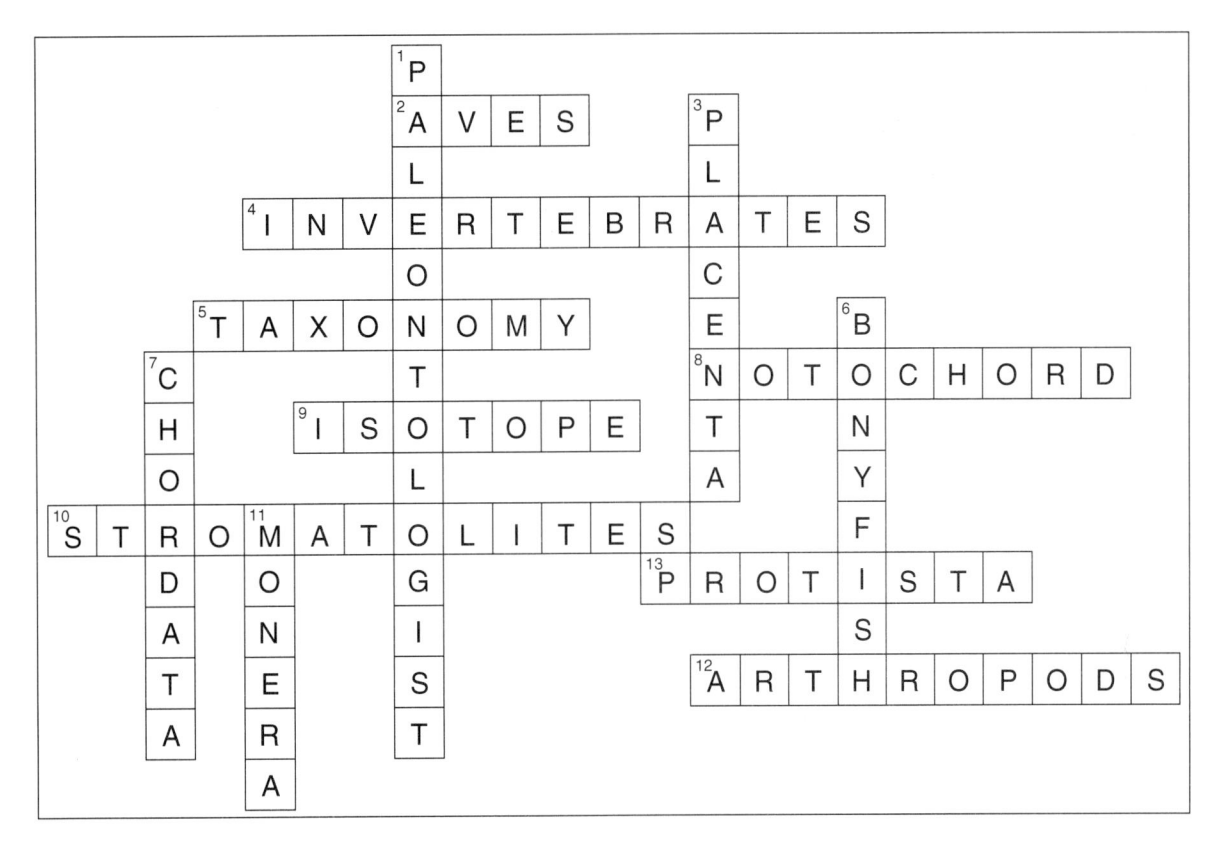

Across

2. The taxonomic classification for birds
4. Animals without backbones
5. The scientific naming of species
8. The embryonic stage of the vertebra
9. An atomic form
10. Layered structures created by algae and bacteria
12. Invertebrates with jointed legs
13. Single-celled animals including algae

Down

1. A scientist who deals with prehistoric life
3. Mammalian organ that nourishes a growing embryo
6. The oldest vertebrate group
7. Animals with a spinal cord
11. Bacterial organisms

CHAPTER 2
Safety

1. For public health concerns, acceptable answers might include:
   (a) the behavior of some animals in a public place (including potential of a bite)
   (b) contamination (dirt, parasites, hair)
   (c) other diners' aversion to animals
   (d) zoonotic potential

   Students may also think of several other reasons, such as "begging for food," knocking over tables and chairs, running into people, dogs fighting with each other, getting in the waste cans, etc. Credit should be given for any reasonable answer.

2. Three observable features of all ticks include:
   (a) They all have eight legs.
   (b) They have the same life cycle.
   (c) They all possess biting mouthparts.

3. Complete the sentences
   (a) intermediate host
   (b) M S D S
   (c) restraint
   (d) reservoir
   (e) Elizabethan collar

4. Crossword Puzzle "Safety"

5. Word Scramble "Safety"

Crossword Safety

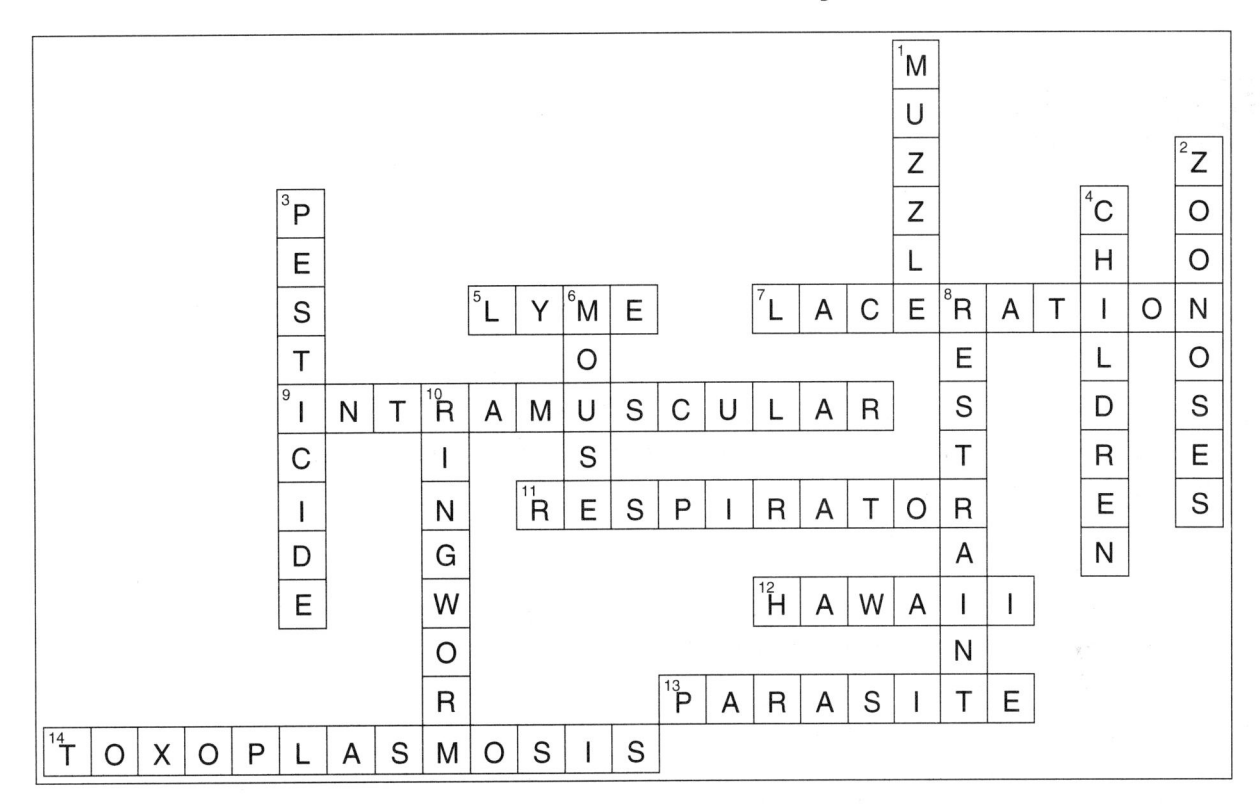

Across

5. Prefix for a disease caused by a tick
7. A wound, a cut or tear in the skin
9. Within the muscle
11. A device that provides protection from inhaling toxic substances
12. The only U.S. state that has never reported a case of rabies
13. An organism that lives on or within another organism
14. A zoonotic disease that can be transmitted from handling contaminated cat litter

Down

1. Device used to prevent an animal from biting
2. Diseases transmitted from animals to humans
3. Chemical that kills insects
4. The primary victims of animal bites
6. Most common intermediate host for *Echinococcus multiloculoaris*
8. The term used for controlling an animal
10. A skin disease caused by a fungus

Word Scramble Safety

1. EFALR — F E R A L
2. RAVLA — L A R V A
3. AOSAMNLLEL — S A L M O N E L L A
4. SEIONLVU — E V U L S I O N
5. UCSNOCECOHCI — E C H I N O C O C C U S
6. SRAADIC — A S C A R I D
7. MYHPN — N Y M P H
8. AITCKISTER — R I C K E T T S I A
9. CETSSNENAU — S U S T E N A N C E

Word Scramble definitions:

1. Feral: a once-domesticated animal that has become wild
2. Larva: the immature state of many species (e.g., ticks)
3. Salmonella: a common bacterial infection transmissible to humans
4. Evulsion: a rip or tear in the skin
5. Echinococcus: tapeworm genus for dogs or cats
6. Ascarid: roundworm
7. Nymph: second stage in the life cycle of ticks
8. Rickettsia: infecctious, zoonotic bacteria carried by ticks
9. Sustenance: food or nourishment

CHAPTER 3
Small Animals as Pets

1. Students are required to support their answers. They should verify the sources they found on the Internet with Web address and date accessed.

   Behavioral concerns (depending on species) include:
   - Basic obedience training
   - Be calm and social in their interactions with strangers and other animals
   - Not react/respond to different stimuli (e.g., smells, crowds, activities, around them) not bark with excitement (dogs)
   - Acceptance of wheelchairs, walking frames, and other medical devices they may encounter during therapy visits

   Health:
   - All animals should have a Health Certificate from a veterinarian stating that the animal is free of external parasites and shows no visible signs of disease.
   - Vaccinations must be current. Vaccinations vary with the species.
   - A record of de-worming, or a negative fecal parasite test.
   - A list of any medications that the pet is prescribed, including heartworm preventives and chemical applications for flea and tick control.
   - The pet should be clean and well groomed, with nails trimmed.

2. This requires a visit to an animal shelter. All responses are dependent on the day of the visit. The adoption application form should be attached to the workbook and confirms that the assignment was completed with an actual visit. Alternatively, if the class is not too large, teachers may consider arranging a group visit. Prearrangements are usually required, and a member of the shelter staff could be available to answer questions.

   Volunteer positions are usually available. Almost all shelters require that new volunteers attend one or possibly more training sessions. Sessions focus on safety for volunteers and the animals, shelter policies, and public relations. Positions may vary, but often include dog-walking and puppy socialization; basic bathing and grooming; cleaning of kennels, cages, and storage areas; and feeding and watering. Experience often leads to a position in pet adoption counseling and the potential of employment. Volunteers are expected to be dependable, on time, and dressed appropriately for the facility and the tasks assigned.

3. A service animal provides assistance to the owner. This can be visual or auditory. Some dogs may be trained to turn lights on and off, retrieve items, and carry packages. A therapy animal visits hospitals, nursing homes, or other institutions and provides comfort to the residents and or patients.

4. The answers will depend on individual research. Students should also be required to understand the medical concerns, not just cite them. For example, a breed may be predisposed to *patellar luxation,* which would not mean much unless the student understood it in the most basic terms, this means "dislocation of the knee joint."

5. This also stresses the importance of understanding medical conditions. The students are asked to build this vocabulary throughout the course. The teacher may wish to review students' lists periodically.

6. Word Search "Small Animals as Pets"

Word Search Small Animals as Pets

W	M	C	X	D	C	F	O	S	X	M	P	J	R	W	X	B	S	Z	L	M	U	B	L	R	X	A
H	Z	W	T	H	E	R	A	P	Y	D	O	G	S	N	N	Q	G	R	I	E	F	A	B	D	U	D
O	O	Y	I	Q	I	I	C	U	O	M	O	H	I	N	Y	G	C	P	R	K	G	S	L	D	O
P	L	S	G	Y	D	N	U	P	W	H	G	U	S	B	E	C	L	Y	A	U	I	R	H	G	A	P
P	B	I	A	Q	W	U	Q	A	D	E	T	M	Z	C	V	A	J	R	V	P	D	Q	C	M	P	T
I	F	J	J	R	Z	O	W	H	X	A	F	A	S	T	A	T	U	S	S	Y	M	B	O	L	S	I
B	I	N	O	U	D	A	V	T	Y	D	C	N	O	F	K	D	R	L	S	U	J	R	D	G	L	O
T	G	M	T	P	H	B	L	Q	E	J	I	E	C	E	C	A	I	S	X	F	A	J	O	I	M	N
I	U	T	W	C	H	W	N	U	U	J	C	S	U	C	D	I	Q	Z	L	X	Y	M	I	N	F
N	T	E	U	T	H	A	N	A	S	I	A	C	O	H	W	S	A	Z	Q	U	H	I	U	E	O	Q	X
N	E	H	N	S	F	R	E	O	F	C	F	C	N	Y	N	E	M	C	U	S	F	B	S	T	E	L
S	C	U	N	A	S	E	N	I	Q	S	I	I	Q	P	S	E	S	H	P	O	N	A	T	E	L	S
E	X	W	Q	L	Q	S	G	O	E	L	U	E	E	N	E	A	U	M	R	O	O	A	I	V	N	W
R	B	T	G	L	D	P	I	N	L	V	T	T	M	E	V	X	E	T	M	Y	K	C	E	G	R
V	L	X	Y	E	Y	O	X	Z	T	E	W	Y	R	O	J	P	U	P	E	Y	I	V	A	Q	R	S
I	Y	K	N	R	X	N	J	L	R	R	M	G	I	G	Q	P	W	D	V	R	L	H	T	A	K	Q
C	L	H	X	G	O	S	M	Y	W	Y	I	D	S	K	Y	C	G	U	E	J	L	T	E	H	A	F
E	S	S	D	I	A	I	P	J	P	D	R	M	U	R	J	E	R	Q	N	B	S	S	D	E	N	Y
D	I	H	A	E	L	B	G	N	E	A	A	O	E	M	C	D	X	A	T	D	H	Y	N	R	E	K
O	R	H	M	S	I	I	A	P	C	U	S	P	K	P	E	C	J	J	R	H	E	E	D	E	S	V
G	E	B	A	A	F	L	Q	I	R	C	F	I	A	A	D	K	X	I	O	Z	L	M	S	D	T	T
S	M	S	G	K	E	I	T	C	O	M	P	A	S	S	I	O	N	C	P	H	T	Q	P	I	H	F
X	U	E	P	B	S	T	P	S	D	R	K	A	G	R	A	G	O	I	E	W	Q	T	E	K
C	M	G	C	A	T	Y	J	J	W	I	N	L	X	G	M	L	I	F	O	T	R	S	I	A	T	W
O	R	Q	Z	K	Y	Q	S	G	U	V	F	L	T	X	A	K	U	M	N	A	U	M	A	R	I	N
M	Y	U	W	R	L	F	S	M	O	E	T	M	Z	A	R	U	M	B	V	C	U	V	Q	Y	Z	R
S	X	U	E	X	E	T	H	Y	P	E	R	T	E	N	S	I	O	N	P	Z	R	Q	T	T	E	Q

Entropion
Spay
No Kill Shelter
Anesthetize
Responsibility
Domesticated

Euthanasia
Sire
Grief
Tumor
Compassion
Lifestyle

Allergies
Dam
Hypertension
Service Dogs
Pedigree
Humane Society

Neuter
Hereditary
Cardiovascular
Therapy Dogs
Status Symbol
Adoption

CHAPTER 4

Animal Rights and Animal Welfare

1. All titles should be preapproved by the teacher. This should not be considered a form of censorship but a concern for appropriateness and relevance.

    The book report may cause concern for a few students, and some may request to watch a video or a DVD instead. If this is acceptable to the teacher, require that all titles be preapproved, or the teacher may wish to provide the video or DVD from the school district's viewing list.

    Teachers should be aware that there are a considerable number of Web sites that address animal rights and animal welfare issues. Many of these provide video footage that is graphic, violent, and disturbing. Many schools have denied access to specific sites; however, it is strongly recommended that if students are allowed Internet access to research some of the issues in this chapter, their activity on the Internet and the sites they access should be carefully monitored for content.

2. This answer will depend on student findings. Additionally, students may wish to collaborate and give a group presentation. Store managers usually welcome students; however, if they are working collectively, it would be a good idea to limit the number of students visiting a retailer at the same time. Teachers may also wish to telephone a select number of stores in advance, explaining the assignment and perhaps even scheduling the student visits.

3. This also requires student research. The textbook is a good place to begin by looking through the chapters where each species is introduced.

4. For some students, this can be a very personal account and they may not wish to share their thoughts and feelings with the class. Other may wish to form debate teams.

5. Word Scramble "Animal Rights and Animal Welfare"

Word Scramble Animal Rights and Animal Welfare

1. PTAE — P E T A
2. NLAAIM HSTIGR — A N I M A L   R I G H T S
3. LMNAAI AERWEFL — A N I M A L   W E L F A R E
4. IROEORESCMTR — E C O T E R R O R I S M
5. MNHUEIAZ — H U M A N I Z E
6. BHRGE — B E R G H
7. ISNVCTOIEVI — V I V I S E C T I O N
8. NLAMIA RAILEOBNTI NROFT — A N I M A L   L I B E R A T I O N   F R O N T
9. CLTNUEIAH — U N E T H I C A L
10. NIITVOR — I N   V I T R O
11. ERISGN — S I N G E R
12. BRASIBT — R A B B I T S
13. EEPCMSISI — S P E C I E I S M
14. COBP — B C O P
15. EVLA — V E A L

CHAPTER 5

Careers in Small Animals

Notes to the teacher:
The majority of the material in this chapter of the workbook is designed to allow students to explore and define their career objectives. All of the exercises are to be completed individually and, as a consequence, all of the responses will be different. In many instances, students will write of the ideal, rather than the practical. This should not be discouraged. It is suggested that grading is based on completion of the projects rather than specific content. The teacher may wish to access the positions selected by the student from the AZA Web site and compare the information recorded.

1. Answers will vary.

2. Answers will vary.

3.
   (a) An externship is a position, paid or unpaid, to further specific educational goals.
       The student is a member of the institution or project, but does not live on the premises.
   (b) A stipend is a specified amount of money given to participants.

5. *-ology*—the science of
   -ist—a qualified person who performs the job: a biolog*ist* studies bi*ology*
   -ician—a qualified person who specializes in one aspect of a science or job

6. Word Scramble "Careers in Small Animals"

75

Word Scramble Careers in Small Animals

1. CSGINEET   G E N E T I C S
2. SRUHNYDBA   H U S B A N D R Y
3. HNIMETRCNE   E N R I C H M E N T
4. MTHPAYE   E M P A T H Y
5. IONNLCG   C L O N I N G
6. LIAOGTPOTSH   P A T H O L O G I S T
7. SLIOITIMOCGRBO   M I C R O B I O L O G I S T
8. YOILYHSPGO   P H Y S I O L O G Y
9. ITNNPISRHE   I N T E R N S H I P

Word Scramble definitions:

1. Genetics: the study of hereditary
2. Husbandry: the care and management of domestic animals
3. Enrichment: providing an environment that stimulates an animal's natural behaviors.
4. Empathy: the ability to understand the feelings of another
5. Cloning: creating an exact genetic duplicate of an animal
6. Pathologist: a scientist who studies disease
7. Microbiologist: a scientist who studies the biology of microscopic life
8. Physiology: the science of understanding body organs and functions
9. Internship: a position that may be either paid or unpaid to further educational goals; similar to an externship, except an intern resides within the facility

CHAPTER 6

Nutrition and Digestive Systems

1. Students will create charts showing the importance of vitamins and then do the same for minerals. The information can be found in the textbook as they read through the material. They may also supplement the charts with information found on the Internet or in specific titles on nutrition.

2. (a) Vitamin A
   (b) Vitamin D

3. Label the different digestive systems.

   Ruminant

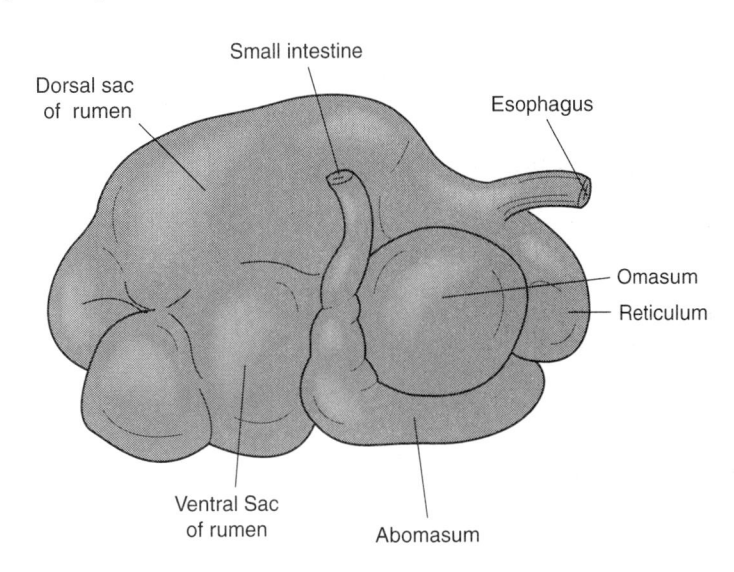

   Dorsal sac of rumen — Small intestine — Esophagus — Omasum — Reticulum — Abomasum — Ventral Sac of rumen

   Monogastric

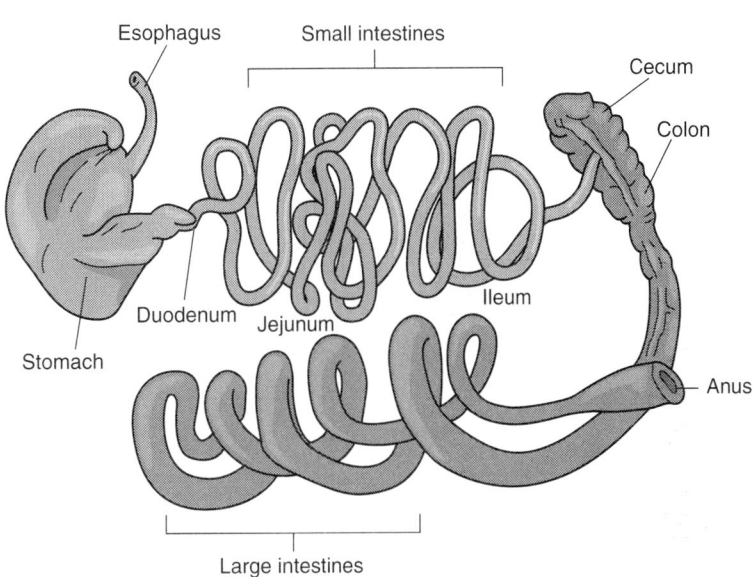

   Esophagus — Small intestines — Cecum — Colon — Stomach — Duodenum — Jejunum — Ileum — Anus — Large intestines

77

Bird

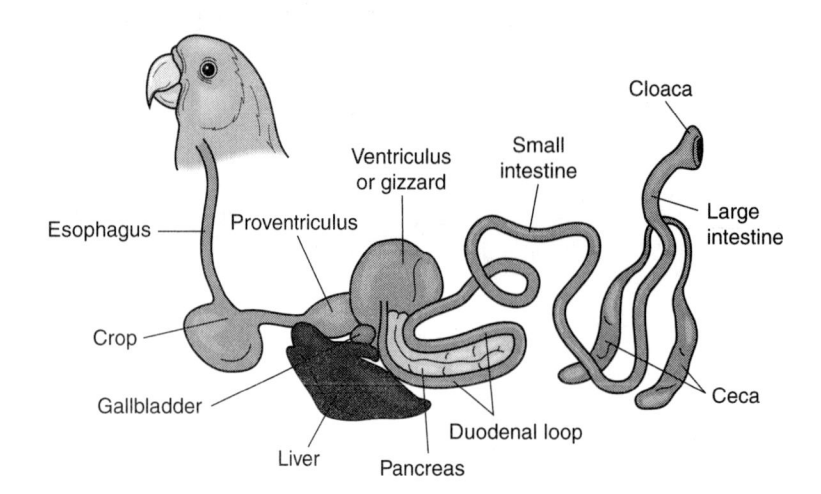

4. Horses and rabbits both possess a large cecum.
5. The cecum contains bacteria that convert roughage to digestible carbohydrates.
6. Crossword Puzzle "Nutrition and Digestion"

Crossword Nutrition and Digestion

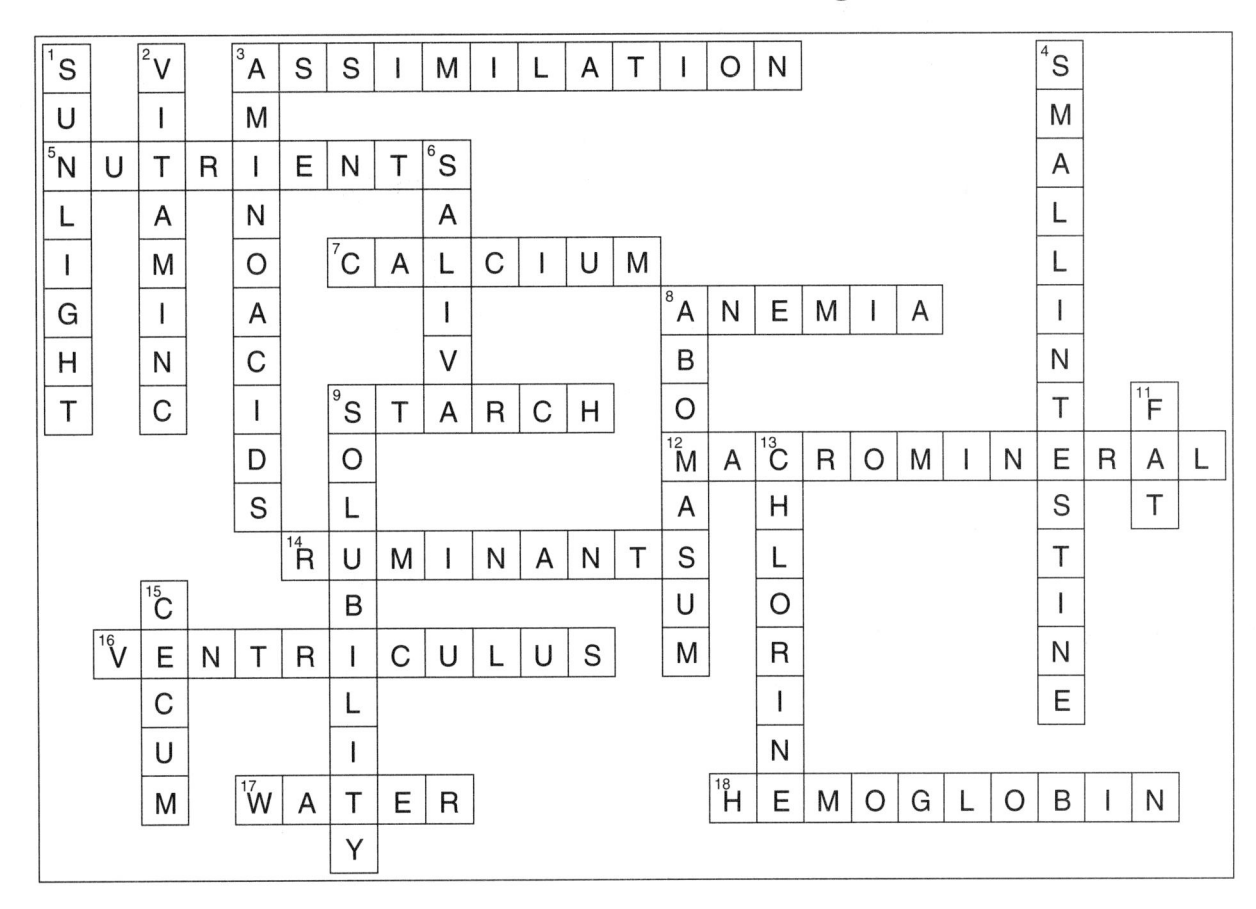

Across

3. Conversion of food so it can be digested
5. Foods that support life
7. Important for bone health
8. Disease associated with an iron deficiency
10. A component of carbohydrates
12. One of the two classifications of minerals
14. Animals with a four-chambered stomach
16. Another term for the gizzard
17. Present in every animal cell
18. Component in blood that carries oxygen

Down

1. Natural source of vitamin D
2. Necessary for the prevention of scurvy
3. The building blocks of proteins
4. Primary site for digestion
6. Contains the first digestive enzymes
8. The "true" stomach in cattle
9. How vitamins are classified
11. Required for the absorption of vitamins A, E, D, and K
13. Salt provides sodium and _____
15. Organ containing bacteria that break down roughage

SECTION 2

CHAPTER 7

Dogs

1. (a) To evaluate the potential behavior of a dog (e.g., Akita vs. German Shepherd)
   (b) To understand how and why the breed was developed and its activity level
   (c) To be able to explain the type of dog to a potential owner
   (d) To recognize the breeds clients may bring in to a clinic or shop
   (e) To evaluate a dog breed that may have genetic issues and concerns

2. The flash cards are valuable in recognizing different breeds by sight. The teacher may also use these for classroom quizzes and "knock-out" competitions.

3. (a) Great Britain
   (b) The needs of the population (agricultural and herding)
   (c) For sport and hunting: hounds, terriers, setters, pointers, spaniels, and retrievers
   (d) Specific utility or working dogs: the Newfoundland, for example, for rescue at sea.

   There are several other reasons for diversity, such a the rise in the standard of living and the development of the toy breeds for companionship.

4. *Female:*
   - No estrus (heat periods) and unwanted male dogs on the premises
   - No unwanted litters
   - Reduces or removes the possibility of mammary gland tumors
   - Removes the possibility of a pyometra (a condition in which the uterus fills up with pus)

   Male:
   - Stops wandering and the potential for being injured (hit by car or dog fight)
   - Removes the potential for testicular cancer and reduces the potential for prostate cancer
   - Reduces mounting behavior in some dogs, although this behavior is also related to dominance and general excitability
   - Often makes dogs less aggressive
   - Less likely to scent-mark perceived territory

5. Word Scramble "Dog Diseases"

6. Crossword Puzzle "Dog Breeds"

Word Scramble Dog Diseases

1. ONUVRSRCIOA   C O R O N A V I R U S
2. RITPOEISSPSOL   L E P T O S P I R O S I S
3. RPUVIOVASR   P A R V O V I R U S
4. MRSDTEIPE   D I S T E M P E R
5. PATSIHETI   H E P A T I T I S
6. HNTEBICIOTSOCHRAR T R A C H E O B R O N C H I T I S
7. NNAAPUAFZIERL   P A R A I N F L U E N Z A

Unscrambled words and the signs of the disease

1. Coronavirus
   (a) vomiting
   (b) diarrhea
   (c) orange, foul smelling feces

2. Leptospirosis
   (a) high fever
   (b) loss of appetite
   (c) reddening of the membranes of the eye and mouth

3. Parvovirus
   (a) vomiting/diarrhea
   (b) yellow/gray, blood-streaked feces
   (c) severe dehydration

4. Distemper
   (a) high fever
   (b) discharge from eyes and nose
   (c) convulsions and neurological damage

5. Hepatitis
   (a) dull, apathetic behavior
   (b) swollen areas: head, neck, lower abdomen
   (c) vomiting and diarrhea

6. Tracheobronchitis
   (a) dry, hacking cough
   (b) gagging
   (c) potential respiratory problems

7. Parainfluenza
   (a) high fever
   (b) dehydration
   (c) difficulty breathing, respiratory tract infection

Crossword Dog Breeds

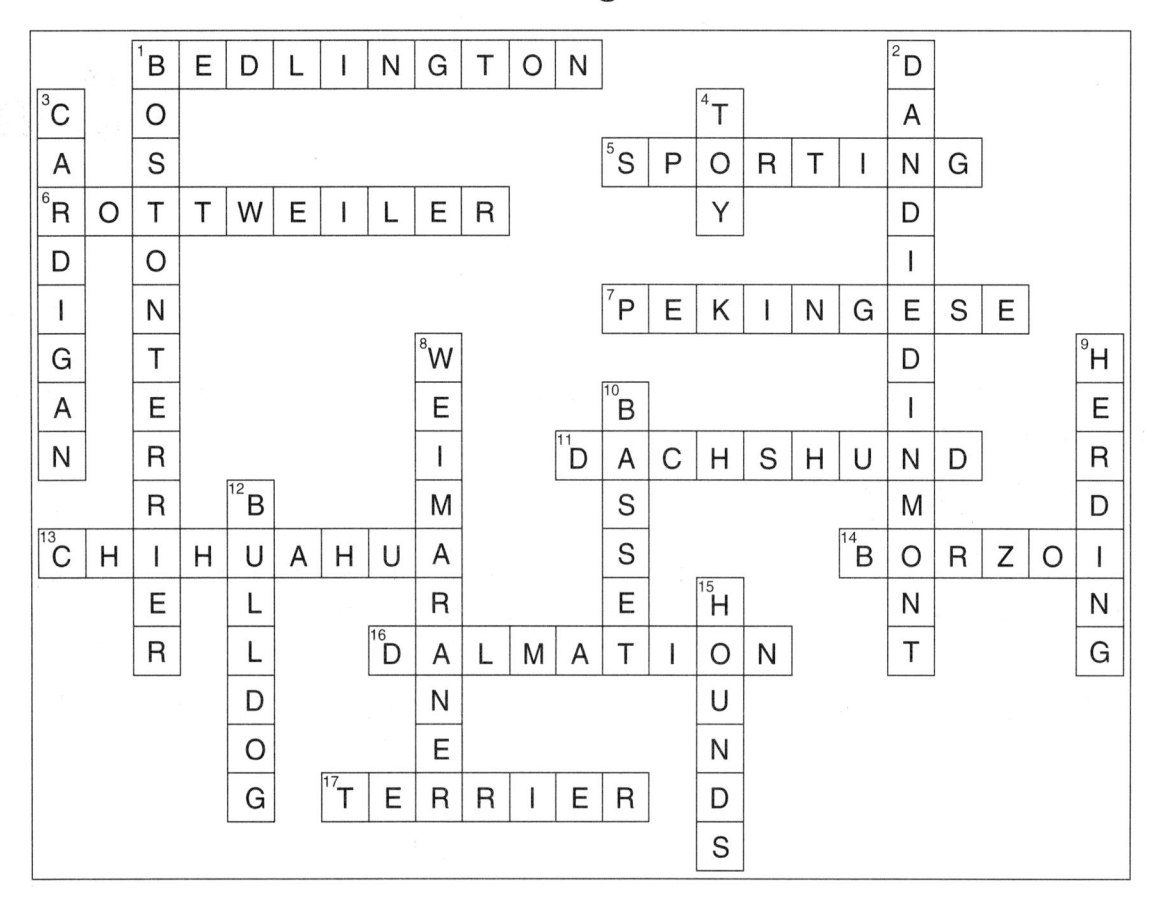

		¹B	E	D	L	I	N	G	T	O	N			²D		
	³C	O							⁴T		A					
	A	S				⁵S	P	O	R	T	I	N	G			
⁶R	O	T	T	W	E	I	L	E	R		Y		D			
D		O										I				
I		N					⁷P	E	K	I	N	G	E	S	E	
G		T			⁸W							D			⁹H	
A		E			E		¹⁰B					I			E	
N		R			I	¹¹D	A	C	H	S	H	U	N	D	R	
		R	¹²B		M		S					M			D	
¹³C	H	I	H	U	A	H	U	A				¹⁴B	O	R	Z	O	I
		E	L		R		E		¹⁵H			O			N	
		R	L		¹⁶D	A	L	M	A	T	I	O	N		G	
			D		N				U							
			O		E				N							
			G	¹⁷T	E	R	R	I	E	R						
									D							
									S							

Across

1. A terrier with the looks of a lamb
5. Group developed to hunt game
6. Descendant of Roman Drover Dogs
7. Once held sacred in China
11. Breed orginally known as the Badgerdog
13. The smallest of all breeds
14. A large sight hound from Russia
16. Often associated with fire trucks and carriages
17. Known for tenacity and going to ground

Down

1. Named after an American city
2. Breed made famous in a novel by Sir Walter Scott
3. A type of corgi
4. The smallest in size but not in affection
8. Breed with a blue-gray coat and golden eyes
9. Group developed to assist with livestock
10. Heavier in bone than any other breed
12. May need to be delivered by caesarean section
15. Some hunt by sight, some hunt by scent

CHAPTER 8
Cats

1. The same method of study used for dog breeds is also useful for cats. The flash cards should be made and used in the same way.

2. (a) Cats require a higher amount of protein
   (b) Cat food has added taurine

3. This is a two-part question. Students should write a short paper discussing the pros and cons of declawing. The paper should also include their personal position on whether or not they would have their own cats declawed.

4. The written paper on a chosen cat breed should include all of the information outlined in the assignment.

5. Word Scramble "Cats"

Word Scramble Cats

1. TCOHLAPIHM — O P H T H A L M I C
2. NMAX — M A N X
3. OMT — T O M
4. JCNOSASOB — J A C O B S O N S
5. ENUQE — Q U E E N
6. AIETNA — T A E N I A
7. FPI — F I P
8. ADNJCUIE — J A U N D I C E
9. ARLEF — F E R A L
10. TKTRESAII — K E R A T I T I S
11. SOMTASLOIXSOP — T O X O P L A S M O S I S
12. ORLTSOMCU — C O L O S T R U M
13. OLCAIC — C A L I C O
14. APLLPAEI — P A P I L L A E
15. ICRIIVUALSC — C A L I C I V I R U S
16. RTUNAIE — T A U R I N E
17. CIITTNATIGN RNEBEAMM — N I C T I T A T I N G   M E M B R A N E
18. AEOCCHL — C O C H L E A

CHAPTER 9
Rabbits

1. Using the metric system:
   (a) The rabbit weighs 1.59 kilograms. For feeding or medication, the weight is rounded to 1.6 kg. (Weight in pounds divided by 2.2 = weight in kg.)
   (b) 1,000 gms/kg
   (c) 1.534 kg (54 oz. divided by 16 = wt. in pounds. Divided by 2.2 = weight in kg)
   (d) 1.50 kg

2. (a) *entero-*: having to do with the intestines
   (b) *toxemia*: tox/tocic emia/blood; a condition of having toxins in the blood
   (c) *gastro-*: having to do with the stomach (gastroenteritis: with a, c, and h "an inflammation of the stomach and intestines")
   (d) *hepatic-*: having to do with the liver (hepatitis, with d and h, "inflammation of the liver")
   (e) *intra-*: within (intramuscular, within the muscle)
   (f) *-osis*: having a condition of a disease (salmonellosis)
   (g) *-cide*: killing, as in insecticide, parasiticide, etc.
   (h) *-itis*: inflammation, keratitis

3. Broad-spectrum antibiotics treat Gm − and Gm + bacteria. The teacher may wish to include a lesson plan to address these bacterial differences.

4. Teeth grinding in many species is a sign of pain.

5. Crossword Puzzle "Rabbits"

Crossword Rabbits

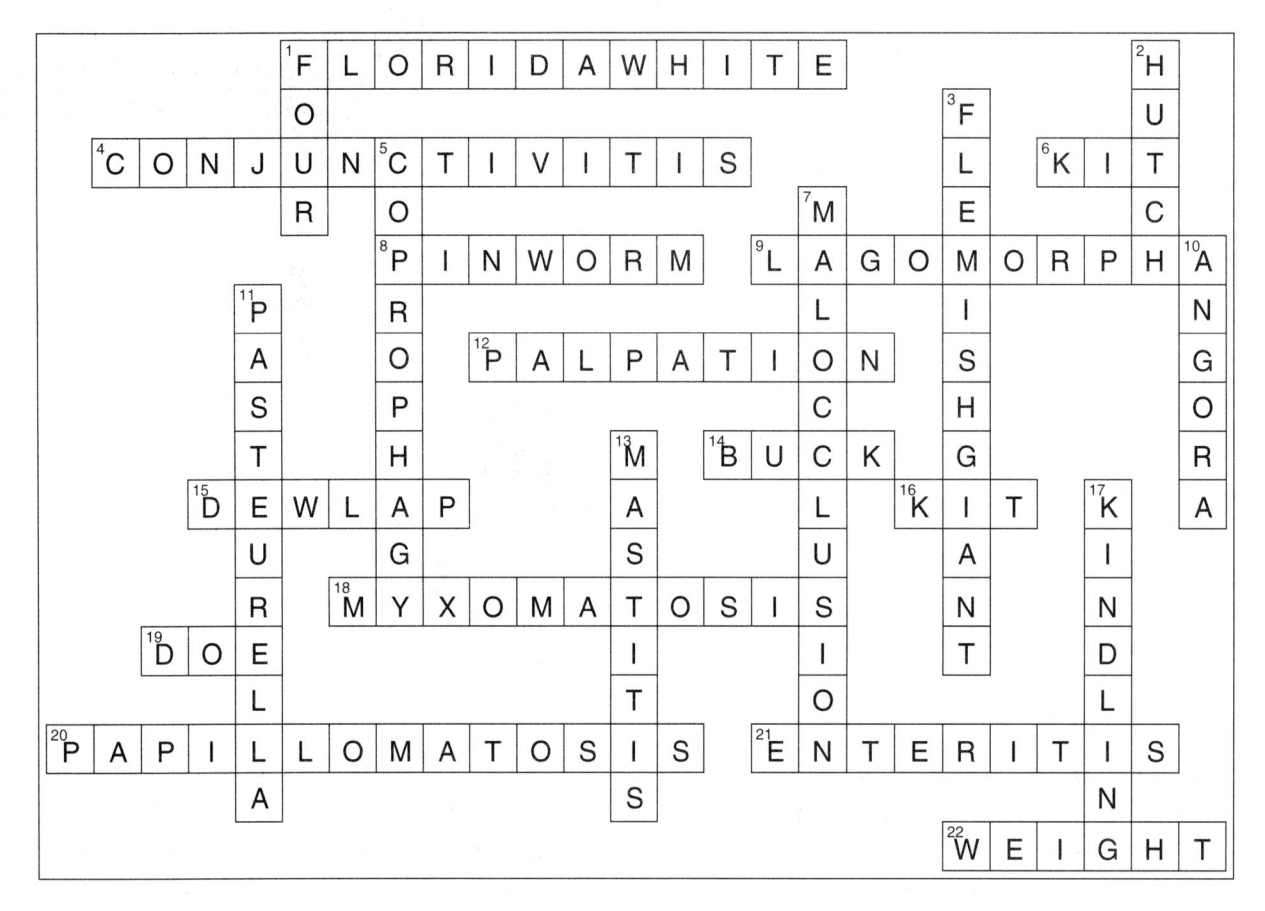

Across

1. Breed most commonly used in research
4. Another term for "Weepy Eye"
6. Term for a young rabbit
8. Common internal parasite of rabbits
9. The taxonomic classification
12. Method used to determine pregnancy
14. Term for a male rabbit
15. The roll of skin under the chin
16. Term for a young rabbit
18. Infectious fatal disease of rabbits caused by a virus
19. Term for a female rabbit
20. Condition of warts around the ears and mouth
21. Inflammation of the intestinal tract
22. How rabbit categories are determined

Down

1. The number of upper incisor teeth
2. Term used for rabbit housing
3. The largest breed of domestic rabbit
5. The consumption of "night feces"
7. The condition of having overgrown teeth
10. Breed with long, soft fur
11. Common bacterial infection of rabbits
13. Inflammation of the mammary glands
17. The term for a rabbit giving birth

CHAPTER 10
Hamsters

1. The basic list of supplies should include the following:
   - Appropriate-sized cage (wire or aquarium with screen lid)
   - Bedding material (purchased by the bag)
   - Exercise wheel
   - Water bottle
   - Food dish
   - Hamster mix feed
   - Hamster

   Answers to a, b, and c will be determined by the above list and costs.

2. The answer will depend on the specific store policy. It is often stated on sales receipts.

3. (a) The likely cause of death is wet tail.
   (b) Wet tail is a highly contagious disease of hamsters. If another hamster is brought home, it too will likely die of this disease. Also, the first hamster has contaminated the entire habitat, and a new, healthy hamster could succumb to the disease and die unless every item is completely disinfected.

4. Students should access this Web site for current updates on LCM. It is a reportable disease, and the CDC (Center for Disease Control) tracks the most recent cases.

5. Word Scramble "Hamsters"

Word Scramble Hamsters

1. SIEMNTGINI — MENINGITIS
2. RFADW — DWARF
3. STEDORUXE — DEXTEROUS
4. CRAATIEB — BACTERIA
5. EHCKE OEPHSUC — CHEEK POUCHES
6. WODO VGSSNIAH — WOOD SHAVINGS
7. ATWRE ETTBLO — WATER BOTTLE
8. UNNLOCRAT — NOCTURNAL
9. TEW TIAL — WET TAIL
10. ECSTN SNLAGD — SCENT GLANDS
11. OINTANIEBRH — HIBERNATION
12. LONOYC — COLONY
13. IANSYR — SYRIAN
14. DEYTD EBAR — TEDDY BEAR
15. AETTVONSII — ESTIVATION
16. CIEDOTDMEC — DEMODECTIC
17. BURHYDNAS — HUSBANDRY
18. ECSXIEER LEHEW — EXERCISE WHEEL
19. RAPPEOLS — PROLAPSE

CHAPTER 11
Gerbils

1. Fill in the blanks
   (a) abdomen flanks
   (b) day diurnal
   (c) gregarious
   (d) monogamous
   (e) mutation
   (f) agouti
   (g) nasal dermatitis
   (h) drum

2. The tail could become caught and the skin of the tail would de-glove.

3. Mice, rats, and hamsters.

4. Cedar contains a toxic oil that causes respiratory problems.

5. Word Scramble "Gerbils"

Word Scramble Gerbils

1. UTNOIMTA — M U T A T I O N
2. OEUGLCAAMF — C A M O U F L A G E
3. RYSETZZ' EAISDES — T Y Z Z E R ' S   D I S E A S E
4. ALIOGONMN — M O N G O L I A N
5. EBYSTIO — O B E S I T Y
6. EARDNTIO — R O D E N T I A
7. UTGOAI — A G O U T I
8. RUSEEZI — S E I Z U R E
9. RNGMUIMD — D R U M M I N G

CHAPTER 12
Rats

1. Students should research the Black Plague, also called bubonic plague, and discuss their findings in class.

2. There have been several outbreaks of bubonic plague in recent years. For the most part, these outbreaks of disease have been confined to national parks in the United States. The fleas in prairie dog colonies and chipmunk populations are responsible for transmitting the disease to the rodent populations.

3. A frightened or defensive rat will:
   (a) Stiffen its body
   (b) Arch its back
   (c) Wag its tail
   (d) Shake
   (e) Erect the hairs on its body

4. Rats can be housed collectively and usually do better with companions. Several females may be housed together or one male with a few females. Two males together are likely to fight. There is no need to remove the male when there is a litter, except to prevent rebreeding.

5. Gerbils are gregarious.

6. Word Scramble "Rats"

Word Scramble Rats

1. EDRTNO — RODENT
2. URTSAT — RATTUS
3. TRNUNLACO — NOCTURNAL
4. IEALG — AGILE
5. YNPITHRRE — PYRETHRIN
6. DOIETSCM TAR — DOMESTIC RAT
7. FASLE — FLEAS
8. P-ANORFGWO — GNAW-PROOF
9. TESANTOGI — GESTATION
10. ODHEOD — HOODED
11. HIEBESKRR — BERKSHIRE
12. NBILOA — ALBINO
13. LAGEUP — PLAGUE
14. NPHOLE — PHENOL
15. CADPE — CAPED
16. RAIRCSRE — CARRIERS
17. EIEGAATRDV — VARIEGATED
18. INNEDRIEGB — INBREEDING
19. PYLOPAXL SUSINPAOL — POLYPLAX SPINULOSA

CHAPTER 13

Mice

1. This question asks the student to compare the four small rodents discussed. Students should create a chart showing their similarities and differences and answer parts a and b of the question.

2. The answer will depend on the information taken from the AFRMA Web site and the student's list of benefits of belonging to a club or an organization.

3. Mouse colonies may contain both males and females, but there will be one dominant male. Only the dominant male will be allowed to breed. Young males may challenge the dominant male, but more often they lead bachelor lives, or if the cage is big enough, they may try to gather younger females and establish other colonies.

4. The most common health problem seen in mice is respiratory disease.

5. The quiz cards should contain only one question and have only one correct answer. They may also include photos of the different varieties of fancy mice.

    Note: the puzzle for this chapter has been replaced with the trivia challenge cards.

CHAPTER 14
Guinea Pigs

1. Guinea pigs are popular as small pets because:
   - They are gentle are rarely bite.
   - The recognize their owners and respond vocally.
   - They are clean and have very little or no odor (if the cage is cleaned).
   - Housing requirements are minimal.
   - They are relatively easy to feed (pelleted foods and some leafy greens).
   - They are easy to handle.

2. Alfalfa is high in protein. Most guinea pig pellets are formulated from alfalfa, so feeding alfalfa hay increases protein consumption. Excess protein can lead to kidney problems. Timothy hay is a better choice.

3. Good sources of vitamin C include:
   - Dandelion greens
   - Beet and carrot tops
   - Kale
   - Timothy hay
   - Citrus fruits
   - Tomatoes

4. Many antibiotics are toxic to guinea pigs. Antibiotics should be given only as prescribed by a veterinarian.

5. After this age a caesarean section will be required to deliver the young and to save the lives of the sow and young. The bones of the pelvis fuse around 7 months of age.

6. Crossword Puzzle "Guinea Pigs"

Crossword Guinea Pigs

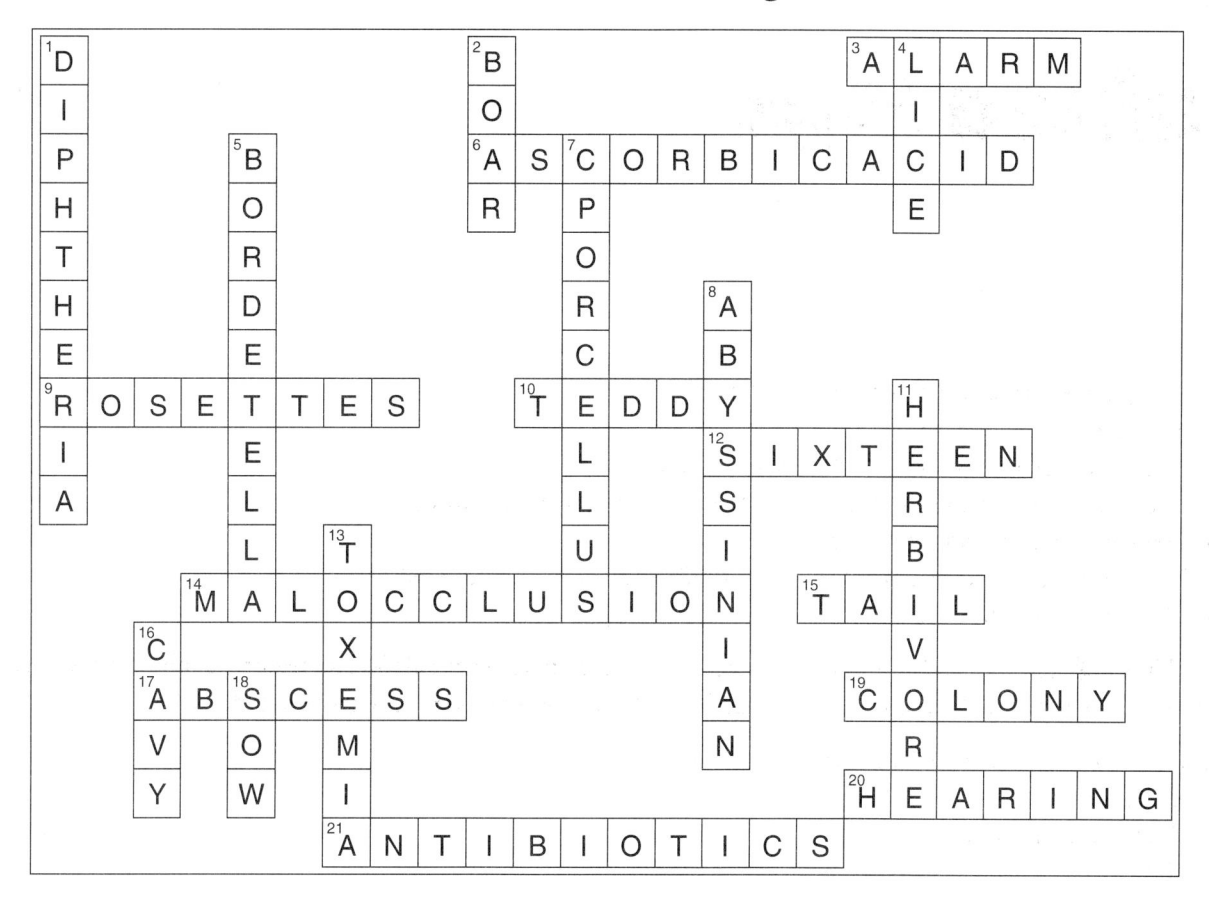

Across

3. One reason guinea pigs chatter their teeth
6. Vitamin C
9. Swirls and cowlicks in the coat
10. Variety with short, kinky fur
12. Number of days in the heat cycle of guinea pigs
14. A "bad bite"
15. Cavies do not have this appendage
17. A swelling caused by the accumulation of pus
19. Collective name for a group of cavies
20. A highly developed sense in guinea pigs
21. Some medicines in this group can cause a toxic reaction

Down

1. Contributed to the discovery and production of this serum
2. Term for a male guinea pig
4. Four of the most common external parasites of guinea pigs
5. Bacteria that also causes kennel cough in dogs
7. Means "little pig" in Latin
8. A variety with a rough, wiry coat
11. Type of diet for guinea pigs
13. A condition that may affect sows in late pregnancy
16. Correct term for a guinea pig
18. Term for a female cavy

CHAPTER 15

Chinchillas

1. There are no population figures available. The chinchilla is listed as critically endangered and may be extinct in the wild.

2. (a) Chinchillas release their fur to escape a predator.
   (b) When threatened or in defense of their young, chinchillas stand upright and spray urine at the source of the threat. Their aim is very accurate.

3. A dust bath provides the opportunity for grooming. Chinchillas roll in the dust to clean their coats. Chinchillas dust is very fine and usually made from powdered pumice.

4. Different coat colors are developed through spontaneous mutations and selective breeding.

5. Enteritis is an inflammation of the intestines. Metritis is an inflammation of the uterus.

6. Word Scramble "Chinchillas"

Word Scramble Chinchillas

1. DRAIAIG   G I A R D I A
2. OTTIZHEROOP   T R O P H O Z O I T E
3. OTZRGEN   G R O T Z E N
4. ITCAHOENPG   P A T H O G E N I C
5. ARPTYMOE   P Y O M E T R A
6. PIIMONCTA   I M P A C T I O N

Word Scramble definitions:

1. Giardia: a microscope intestinal protozoan that causes diarrhea; has zoonotic potential
2. Tropohozoite: the growing stage of a protozoan parasite
3. Grotzen: the center back strip of a chinchilla pelt
4. Pathogen: any disease-causing organism
5. Pyometra: an accumulation of pus within the uterus

CHAPTER 16

Ferrets

1. Students are required to check the local laws regarding ferrets. They may be legal to keep in a state, but prohibited in towns or communities. Just because ferrets are offered for sale doesn't mean they are legal. Retailers have had ferrets confiscated and destroyed because of a local ordinance that prohibits ferrets as pets. If ferrets are illegal as pets, the answers may vary, but often-stated reasons are:
   - Ferrets attack children and other pets.
   - Ferrets are carriers of rabies.
   - If ferrets got loose, or were intentionally released, they could establish breeding populations.
   - Hunting with ferrets is illegal in the United States.

   The teacher may wish to follow through with these reasons: are the reasons valid or based on inaccurate information and hearsay about pet ferrets?

   In areas where ferrets are legal, owners may have to pay a licensing fee. Many areas permit pet ferrets but require that they be spayed or neutered. It is usually mandatory that ferrets receive annual rabies vaccinations.

2. The majority of pet ferrets sold in the United States are bred by Marshall Farms in New York.

3. The two blue tattoos in one ear of a ferret signify that it was bred by Marshall Farms and has already been spayed or neutered and descented.

4. When ferrets play, they can become very rough with one another. When initiating play they hunch their backs, skitter backward "bounce" up and down, shake their heads, and sometimes hiss or "chortle." This is not to be mistaken for an aggressive attack.

5. Ferrets are induced ovulators. Jills remain in heat until they are bred. Estrogen is produced in excess, which causes aplastic anemia. Jills must be either bred or spayed. Hormones may be used to bring the jill out of estrus, but this is not recommended as an ongoing therapy.

6. Crossword Puzzle "Ferrets"

Crossword Ferrets

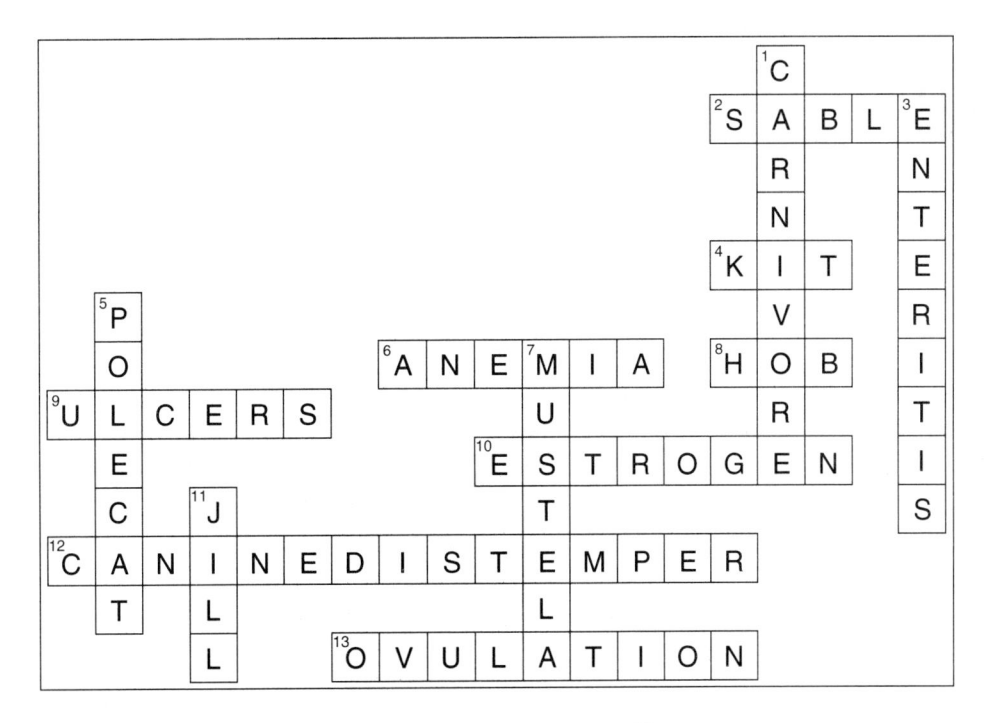

Across

2. The most common ferret color
4. Term for a young ferret
6. Severe condition due to the excess production of estrogen
8. Term for a male ferret
9. Stomach problem caused by *Helicobacter*
10. Female hormone
12. Vaccine for ferrets to prevent fatal viral disease
13. The release of eggs from the ovaries

Down

1. Diet fed to ferrets
3. Inflammation of the intestines
5. European relative to the ferret, used in hunting rabbits
7. Genus that also includes weasels, minks, and polecats
11. Term for a female ferret

CHAPTER 17
Amphibians

1. Students are requested to draw the life cycle of an amphibian in the space provided. They should not refer to the textbook. Regardless of the species, all amphibians have the same life cycle. It is represented in Figure 17–25.

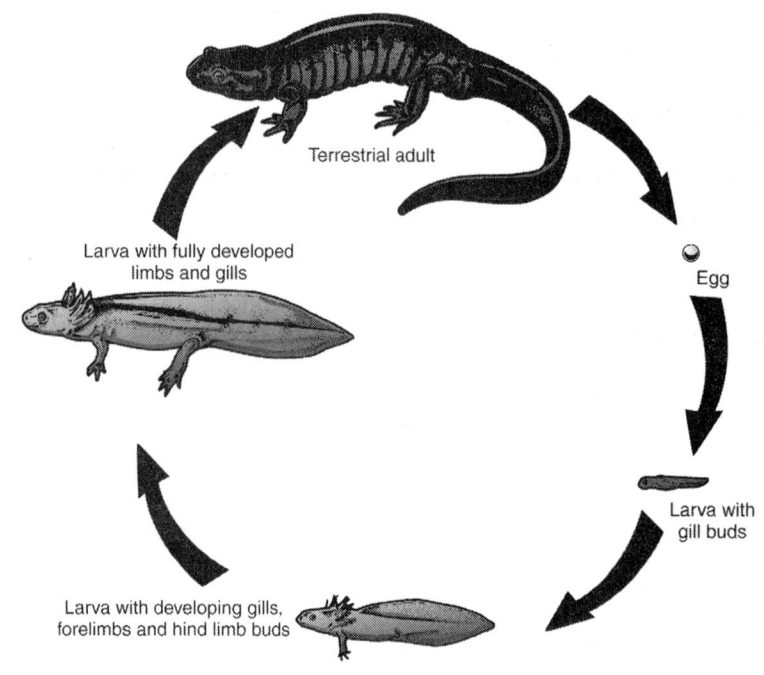

2. (a) Bare hands can disrupt the delicate mucous coating on the skin of some amphibians; they can also absorb chemicals from human hands (soap, lotions, etc.).
   (b) Some amphibians produce toxins from glands in the skin. An easy way to remember to wear gloves is "Toxins in, toxins out."

3. (a) The African clawed frog is completely aquatic.
   (b) The Tiger Salamander should have a woodland habitat (with access to a water bowl).
   (c) Leopard frogs should have a semiaquatic habitat.

4. (a) Clean water is essential for the health of amphibians to prevent bacterial infections from contaminated water (red leg is an example).
   (b) Amphibians "breathe" through their skin and absorb the chemicals in water that has not been dechlorinated. Tap water contains enough chlorine to be toxic to amphibians.

5. Aquatic frogs are the most affected by red leg. It is caused by *aeromonas* bacteria that thrive in contaminated, dirty water. It is best prevented by providing clean, dechlorinated water with a good filtration system.

6. Word Scramble "Amphibians"

Word Scramble Amphibians

1. UTACADE — C A U D A T E
2. RANOUEOSM — A E R O M O N U S
3. SPEOAPRORHTME — S P E R M A T O P H O R E
4. SOSSOIM — O S M O S I S
5. ARALV — L A R V A
6. SEAATILIN — S A L I E N T I A

Word Scramble definitions:

1. Caudate: amphibian with a tail
2. *Aeromonas:* bacterium that causes red leg
3. Spermatophore: a sperm packet deposited by the male
4. Osmosis: the flow of a liquid from a higher concentration to a lower concentration through a semipermeable membrane
5. Larva:
6. Salienta: taxonomic genus that includes frogs and toads

CHAPTER 18

Reptiles

1. The reptile heart is three chambered, the mammalian heart is four chambered. This anatomical difference has the effect of slowing metabolic activity.

2. Turtles are aquatic or semiaquatic, whereas tortoises live on land. Turtles are omnivores and tortoises are herbivores. Most species of tortoises are larger than turtles.

3. UBV lighting is necessary for:
   (a) vitamin D synthesis
   (b) prevention of metabolic bone disease
   (c) calcium metabolism
   Another correct answer: formation of pigment.

4. The snake could be placed in a damp burlap bag or the humidity of the enclosure could be increased. Alternatively, it could be soaked in warm water and the shed could be assisted by gently rolling the dead skin in the direction of the scales, moving from head to tail. Great care must be taken with the eye caps.

5. Knowing the POTZ range for a specific species can help to provide the healthiest and most natural environment with regard to temperature gradients and humidity levels.

6. Crossword Puzzle "Reptiles"

Crossword Reptiles

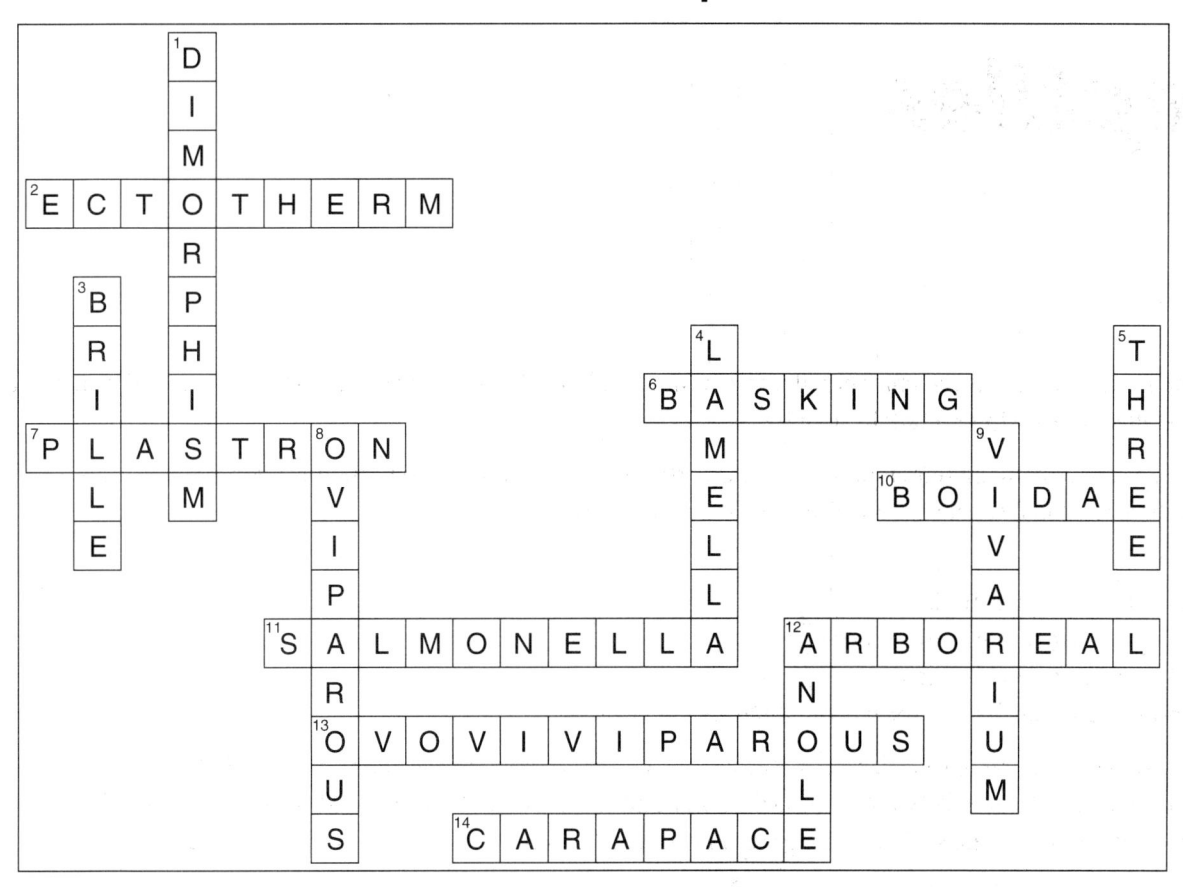

			¹D														
			I														
			M														
	²E	C	T	O	T	H	E	R	M								
			R														
	³B		P					⁴L						⁵T			
	R		H				⁶B	A	S	K	I	N	G				
	I		I					M				⁹V		H			
⁷P	L	A	S	T	R	⁸O	N				¹⁰B	O	I	D	A	E
	L			M		V		E				V		E			
	E					I		L				A					
						P		L		¹²A							
				¹¹S| A	L	M	O	N	E	L	L	A	R	B	O	R	E	A	L
				R						N				I			
				¹³O	V	O	V	I	V	I	P	A	R	O	U	S
				U						L				M			
				S		¹⁴C	A	R	A	P	A	C	E			

Across

2. Correct term for "cold-blooded"
6. What reptiles do to raise their body temperature
7. The lower part of a chelonian shell
10. Taxonomic family with the largest species of snakes
11. Bacterial disease that can be transmitted to humans
12. Living in trees
13. Eggs that hatch internally
14. The upper part of a turtle shell

Down

1. Visual differences between the senses
3. The transparent eye covering
4. The adhesive pads on the feet of geckos
5. Number of chambers in a reptile heart
8. Eggs that hatch externally
9. A habitat that replicates a natural environment
12. The smallest member of the iguana family

CHAPTER 19

Birds

1. Anatomical features that give a bird the ability to fly:
   (a) hollow bones
   (b) air sacs within the wings and body
   (c) sternum (keel bone)
   (d) fused ribs for stability
   (e) strong pectoral muscles attached to the wings
   (f) feathers

2. Students should draw the differences between the beaks and feet of two major bird groups, psittacines and passerines. Psittacines have a hooked bill (the parrots) and four toes, two pointed forward and two point backward. Passerines (perching birds) also have four toes, but three are forward and one is back. The bill is straight.

3. Label the anatomical parts of the female reproductive system.

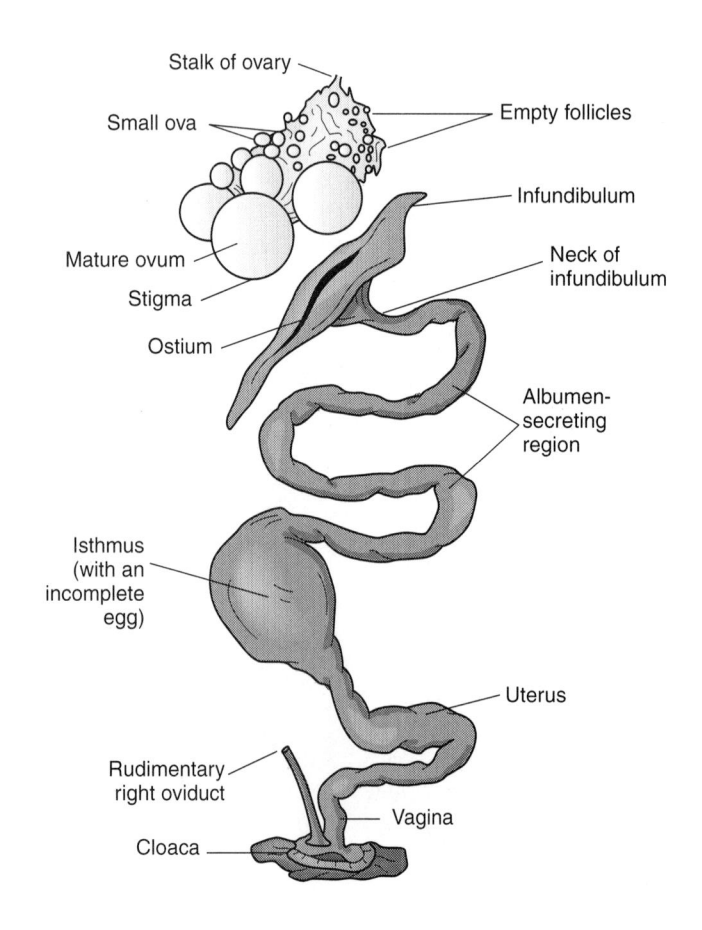

Labels: Stalk of ovary, Small ova, Empty follicles, Infundibulum, Mature ovum, Neck of infundibulum, Stigma, Ostium, Albumen-secreting region, Isthmus (with an incomplete egg), Uterus, Rudimentary right oviduct, Vagina, Cloaca

4. The tongue of a lorry has bristles, not unlike a toothbrush. They are nectar feeders and use the tongue to lap nectar.

5. The disease is psittacosis. Students should differentiate between the terms *psittacosis* (the disease in birds) and *chlamydiosis* (the same disease in humans). The discussion should also include the causative bacterial agent (*C. psittaci*) and the methods of transmission to people.

6. Crossword "Puzzle Birds"

Crossword Birds

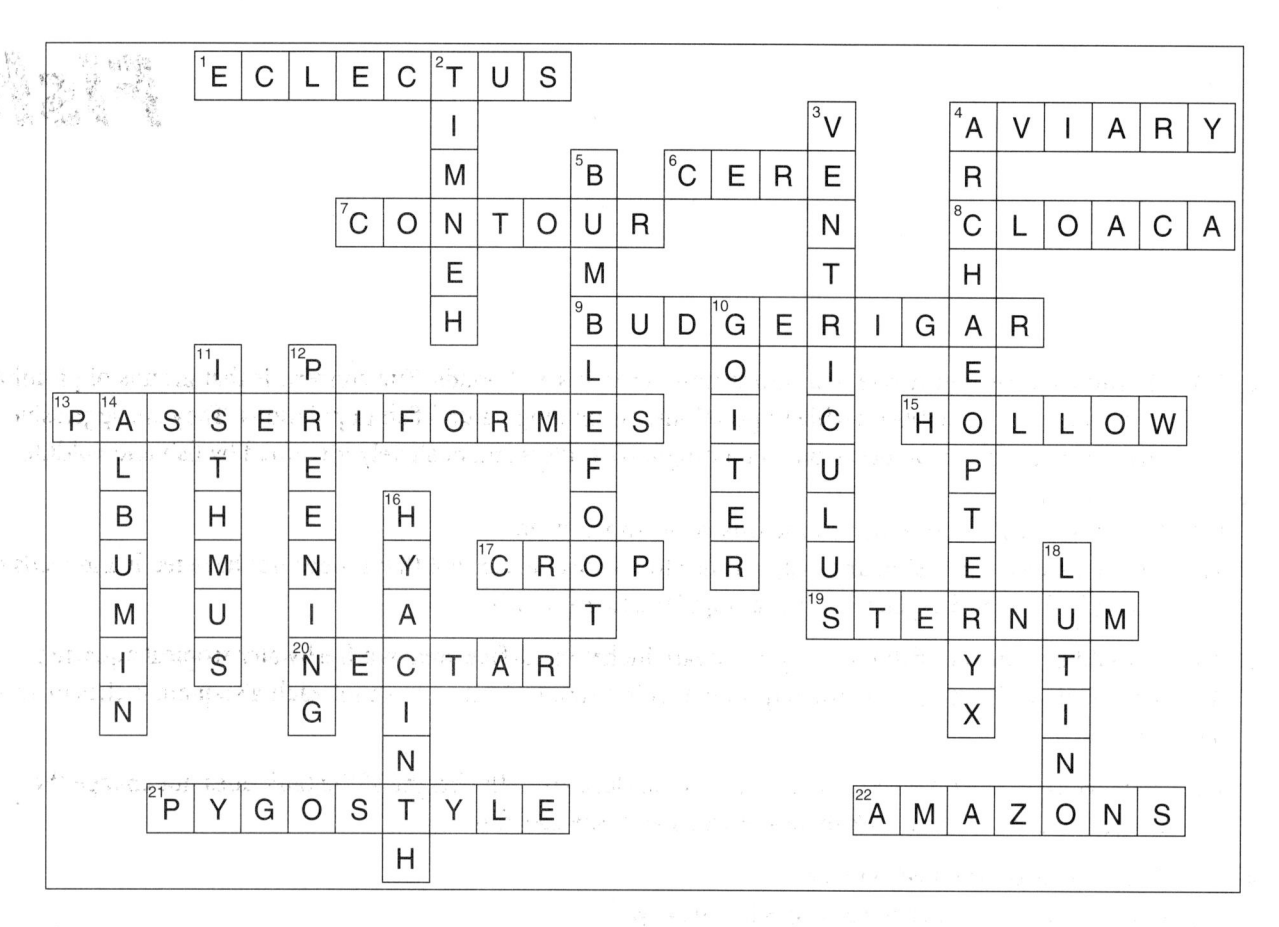

Across

1. Sexually dimorphic bird with male and female having different colors
4. A place where birds are housed
6. The fleshy part of a bird's beak
7. The feathers that determine the shape of the bird
8. Junction of the urinary, digestive, and reproductive tracts
9. Popular pet bird often called a parakeet
13. The perching birds
15. Describing the bones of birds
17. An enlargement of the esophagus that stores food
19. The keel bone
20. A food that needs to be provided for lories
21. The last section of the spine, the tail area
22. The largest genus group

Down

2. A subspecies of the African Gray
3. The largest organ in the digestive system of a bird
4. The first bird
5. A painful ailment of the feet
10. A condition caused by an iodine deficiency
11. Anatomical part where the shell is formed
12. When birds groom themselves
14. The thick, white substance secreted around the yolk
16. The largest of all the macaws
18. A cockatiel that is primarily yellow

CHAPTER 20
Fish

1. (a) Piranhas are frequently released into freshwater rivers and ponds. The concern is that groups of piranhas could school and establish viable populations, decimating natural fish populations. They are aggressive carnivores and have, on occasion, been caught by anglers and positively identified by fish and wildlife officers.
   (b) The student will need to determine this yes-or-no answer.
   (c) The Red Pacu is very similar in appearance to the piranha, except that it does not have teeth and feeds on small plankton and vegetation. It is a popular aquarium fish.

2. There should be 1 inch of fish for every 30 square inches of surface area in a freshwater tropical aquarium. If it is a marine (saltwater) tank, the requirements increase to 1 inch of fish for each 48 square inches of surface area.

3. Length × width × height = 432 square inches of surface area. The height of the tank does not change the surface area. 432 divided by = 43 inches of tropical freshwater fish.

4. (a) The instrument is a hydrometer.
   (b) Other terms: water density and specific gravity.

5. Students should label the anatomy of a fish as shown in Figure 20-3.

6. Crossword Puzzle "Fish"

Crossword Fish

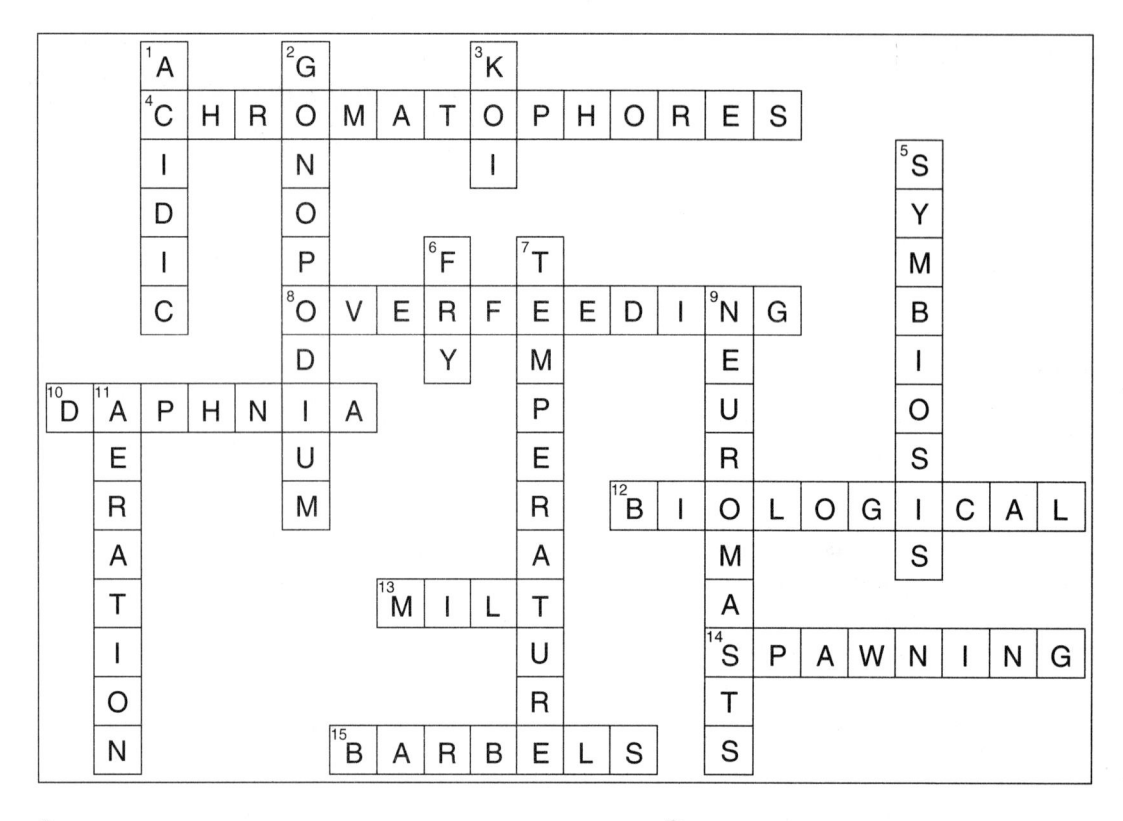

Across

4. Pigment cells that give fish their color
8. Contributes to an increase of ammonia and nitrates
10. Small crustaceans called "water fleas"
12. Filtration system that uses bacteria
13. Fish sperm
14. Reproduction ritual for egg-layers
15. Whisker-like projections around the mouth

Down

1. pH of 0
2. The modified anal fin of some male fish
3. An ornamental species of carp from Japan
5. A mutually beneficial relationship
6. Term for newly hatched or newborn fish
7. A critical factor in most tanks
9. The pressure-sensitive cells within the lateral line
11. The exchange of oxygen and carbon dioxide in the water

CHAPTER 21

Hedgehogs

1. Fad pets often become popular because of films or other media. Consider Beatrix Potter and her tales of Mrs. Tiggywinkle. There are also cartoon strips that feature hedgehogs, the Sonic the Hedgeho video game, and other games and media that popularize species. The problem arises when owner expectations clash with reality and the new "fad pet" is no longer wanted.

2. Hedgehogs are relatively recent as pets, and some of their medical concerns are still being researched. Wobbly hedgehog syndrome is one of the concerns about which little is really known, other than that is becoming an increasing medical concern in pet hedgehogs. The cause is unknown.
   It is most likely a neuromuscular disease. Early signs of this condition are collapse of the hind limbs followed by hind-limb paralysis. The paralysis progress to the front legs and all mobility is lost.
   The eventual outcome is death. For humane reasons, the hedgehog should be euthanized once a diagnosis is confirmed.

3. The student needs to consult with a veterinarian who regularly sees hedgehogs. The most common treatment is ivermectin, which is given by injection. Depending on the severity of the infestation, one or more treatments may be required.

4. The usual source of a mite infestation is the bedding material or natural wood that is added to the habitat.

5. Word Scramble "Hedgehogs"

Word Scramble Hedgehogs

1. DARAIRHE   D I A R R H E A
2. ITRIAPRTONU   P A R T U R I T I O N
3. SNSPIE   S P I N E S
4. NERVIOCESTI   I N S E C T I V O R E
5. GONITNANI   A N O I N T I N G
6. IKNTRAE   K E R A T I N
7. THLOGE   H O G L E T
8. .A NEVRUASILBT   A . a l b i v e n t r u s

Word Scramble definitions:

1. Diarrhea: abnormally loose stool, liquid feces
2. Parturition: the process of giving birth
3. Spines: the sharp, outer covering of the hedgehog's body
4. Keratin: the hard substance that makes up hair, horns, and spines
5. Hoglet: term for a young hedgehog
6. *A. albiventrus*: the genus and species of the hedgehogs most commonly kept as pets

CHAPTER 22
Sugar Gliders

1. Male sugar gliders have a large scent gland on the top of the head. This becomes more pronounces as the sugar glider matures.

2. The answer should be evident from the photograph.

3. The student will need to check with animal control regarding the legality of keeping sugar gliders.

    Responses vary depending on the locality. A common reason is simply, "they're wild animals."

4. *Giardia*, a protozal parasite, has been reported in humans with pet sugar gliders.

5. The koala is one example of a mammal with syndactylism. What it has in common with the sugar glider is that they are both arboreal marsupials and they both are native to Australia.

6. Word Scramble "Sugar Gliders"

Word Scramble Sugar Gliders

1. OCAACL — C L O A C A
2. MUG — G U M
3. SECNT NDALG — S C E N T G L A N D
4. NYASDLTCY — S Y N D A C T Y L
5. ONYCSIMCAITSO — A C T I N O M Y C O S I S
6. URNOANCLT — N O C T U R N A L
7. SPA — S A P
8. SARUPIMUM — M A R S U P I U M
9. IAACAC — A C A C I A
10. RIPLSMAAU — M A R S U P I A L
11. MGPUAITA — P A T A G I U M
12. UHCOP — P O U C H
13. EMNIOVORS — O M N I V O R E S
14. IRSAMIPP — P R I A P I S M
15. JESOY — J O E Y S